Network Science with Python

Explore the networks around us using network science, social
network analysis, and machine learning

David Knickerbocker

BIRMINGHAM—MUMBAI

Network Science with Python

Group Product Manager: Gebin George

Publishing Product Manager: Ali Abidi

Senior Editor: David Sugarman

Content Development Editor: Priyanka Soam

Technical Editor: Rahul Limbachaya

Copy Editor: Safis Editing

Project Coordinator: Kirti Pisat

Proofreader: Safis Editing

Indexer: Hemangini Bari

Production Designer: Jyoti Chauhan

Marketing Coordinator: Shifa Ansari and Vinishka Kalra

First published: February 2023

Production reference: 1240223

Published by Packt Publishing Ltd.
Livery Place
35 Livery Street
Birmingham
B3 2PB, UK.

ISBN 978-1-80107-369-1

www.packtpub.com

Acknowledgements

Writing a book is a challenge, and I have had so much support along the way. I would like to thank several people and groups.

First, I would like to thank my family for constantly supporting me throughout my life and career. With regard to this book, I would especially like to thank my daughter, who helped with the editing process.

Next, I want to thank the Packt teammates I worked with on the book. Thank you for being so patient with me and for helping me at each step. It was a joy to work with you all. Thanks for all of the editing and guidance.

I would also like to thank my friends, who encouraged me to write this book and encouraged me to keep going at each step in the process. I would especially like to thank Ravit Jain. He has done so much to bring the data science community together, and he was the one who finally persuaded me to get over my fears and be bold and write my first book.

And finally, I want to thank you, dear readers, for reading my first book. Thank you for taking the time to read my thoughts and ideas. Thank you for learning with my implementations of code. Thank you for joining me on this learning adventure. I hope this book helps you.

David Knickerbocker

Contributors

About the author

David Knickerbocker is the chief engineer and co-founder of VAST-OSINT. He has over 2 decades of rich experience working with and around data in his career, focusing on data science, data engineering, software development, and cybersecurity. During his free time, David loves to use data science for creative endeavors. He also enjoys hanging out with his family and cats. He loves to read data science books outside, in the sun.

About the reviewers

Shashi Prakash Tripathi is a highly skilled senior NLP engineer at Tata Consultancy Services' Analytics and Insights unit in Pune, India. With over 6 years of industry experience, he is adept at transforming complex problems into successful minimum viable products. His areas of expertise include machine learning, deep learning, information retrieval, natural language processing, social network analysis, and data science. Tripathi is an IEEE professional member who has published numerous articles in reputable journals and conferences such as IEEE and Springer. He also serves as a guest instructor at Liverpool John Moores University.

Albert Yumol is a senior data scientist based in the Netherlands working on financial modeling and credit analytics aided by artificial intelligence. He got his nickname by being addicted to scripting automation and all things open source. Together with various tech communities, he organizes platforms for learning human-centered tech skills. He believes that technology should be accessible to all regardless of race, age, gender, and social status. As such, he works as an activist helping governments and organizations become data literate and use it as a tool for social development and transformation. He is active on LinkedIn, engaging more than 20,000 followers who inspire him to contribute to defining the future of data workers.

Sougata Pal is a passionate technology specialist playing the role of an enterprise architect in software architecture design and application scalability management, team building, and management. With over 15 years of experience, he has worked with different start-ups and large-scale enterprises to develop their business application infrastructure to enhance customer reach. He has contributed to different open source projects on GitHub to empower the open source community. For the last couple of years, Sougata has played around with federated learning and cybersecurity algorithms to enhance the performance of cybersecurity processes by federated learning concepts.

Table of Contents

2

Network Analysis   39

3

Useful Python Libraries   65

Part 2: Graph Construction and Cleanup

4

NLP and Network Synergy   89

5

Even Easier Scraping!   149

6

Graph Construction and Cleaning   181

Part 3: Network Science and Social Network Analysis

7

Whole Network Analysis   211

10

Supervised Machine Learning on Network Data   311

11

Unsupervised Machine Learning on Network Data   335

Preface

Networks are everywhere and in everything. Once you learn to see them, you will find them everywhere. Natural data is not structured. It is unstructured. Language is one example of unstructured data, and networks are another example. We are, at all times, being influenced by the things that exist around us. I wrote this book in the hope of showing how language data can be transformed into networks that can be analyzed and then showing how the analysis itself can be implemented.

This book does not intend to teach every aspect of network science but to give enough of an introduction so that the tools and techniques may be used confidently to uncover a new branch of insights.

Who this book is for

My intention for this book was for it to serve as a bridge between computer science and the social sciences. I want both sides to benefit, not just one. My hopes are that software engineers will learn useful tools and techniques from the social sciences, as I have, and also that those in the social sciences will be able to scale up their ideas and research using software engineering. Data scientists will also find this book useful for going deeper into network analysis.

However, this book will not just be useful to social scientists, software engineers, and data scientists. Influence happens all around us. Understanding how to measure influence or map the flow of ideas can be very useful in many fields.

What this book covers

Chapter 1, *Introducing Natural Language Processing*, introduces **natural language processing** (NLP) and how language and networks relate. We will discuss what NLP is, where it can be useful, and walk through several use cases. By the end of this chapter, the reader will have a basic understanding of NLP, what it is, what it looks like, and how to get started in using NLP to investigate data.

Chapter 2, *Network Analysis*, introduces network analysis, which is useful for analyzing aspects of a network, such as its overall structure, key nodes, communities, and components. We will discuss different types of networks and different use cases, and we will walk through a simple workflow for collecting data for network research. This chapter gives a gentle introduction to understanding networks and how understanding networks may be useful in solving problems.

Chapter 3, Useful Python Libraries, discusses the Python libraries used in this book, including where to go for installation instructions and more. This chapter also provides some starter code to start using libraries and to check that they are working. Libraries are grouped into categories so that they can be easily understood and compared. We will discuss Python libraries for data analysis and processing, data visualization, NLP, network analysis and visualization, and **machine learning** (**ML**). This book does not depend on a graph database, reducing the overhead of learning network analysis skills.

Chapter 4, NLP and Network Synergy, discusses how to take text data and convert it into networks that can be analyzed. We will learn to load text data, extract entities (people, places, organizations, and more), and construct social networks using text alone. This allows us to create visual maps of how characters and people interact in a body of text, and these can serve as a way to interactively gain a richer contextual awareness of the content or entities being studied.

In this chapter, we will also discuss crawling, also known as web scraping. Learning how to collect data from the internet and use it in NLP and network analysis will supercharge an individual's ability to learn about either or both. It also unlocks the ability to research things that interest you rather than using someone else's datasets.

Finally, we will convert the text into an actual network and visualize it. This is a very long chapter because it covers each of the steps involved in converting text into networks.

Chapter 5, Even Easier Scraping, shows even easier methods for collecting text data from the internet. Certain Python libraries have been created for pulling text data from news sites, blog websites, and social media. This chapter will show you how to easily and quickly get clean text data for use in downstream processes. In this chapter, we will convert text into an actual network, and then we will visualize the network.

Chapter 6, Graph Construction and Cleaning, dives into working with networks. The chapter starts by showing how to create a graph using an edge list and then describes important concepts such as nodes and edges. Over the course of the chapter, we will learn to add, edit, and remove nodes and edges from graphs, which is all a part of graph construction and cleaning. We will conclude the chapter with a simulation of a network attack, showing the catastrophically destructive effect that removing even a few key nodes can have on a network. This shows the importance that individual nodes have on an entire network.

Chapter 7, Whole Network Analysis, is where we really get started with network analysis. For instance, we will look for answers on the size, complexity, and structure of the network. We will look for influential and important nodes and use connected components to identify structures that exist in the network. We will also briefly discuss community detection, which is covered in much greater depth in *Chapter 9*.

Chapter 8, Egocentric Network Analysis, investigates the egocentric networks that exist in a network. This means that we will look closer at nodes of interest (ego nodes) and the nodes that surround them (alters). The goal is to understand the social fabric that exists around an ego node. This can also be useful for recognizing the communities that an individual belongs to.

Chapter 9, Community Detection, discusses several approaches to identifying the communities that exist in a network. Several powerful algorithms for community detection will be shown and discussed. We will also discuss how connected components can be used to identify communities.

Chapter 10, Supervised Machine Learning on Network Data, shows how we can use networks to create training data for ML classification tasks. In this chapter, we will create training data by hand, extracting useful features from a network. We will then combine those features into training data and build a model for classification.

Chapter 11, Unsupervised Machine Learning on Network Data, shows how unsupervised ML can be useful for creating node embeddings that can be used for classification. In the previous chapter, we created network features by hand. In this chapter, we will show how to create node embeddings using Karate Club and then use them for classification. We will also show how to use Karate Club's GraphML models for community detection.

Download the example code files

You can download the example code files for this book from GitHub at `https://github.com/ PacktPublishing/Network-Science-with-Python`. If there's an update to the code, it will be updated in the GitHub repository.

We also have other code bundles from our rich catalog of books and videos available at `https:// github.com/PacktPublishing/`. Check them out!

Conventions used

There are a number of text conventions used throughout this book.

`Code in text`: Indicates code words in text, database table names, folder names, filenames, file extensions, pathnames, dummy URLs, user input, and Twitter handles. Here is an example: "Mount the downloaded `WebStorm-10*.dmg` disk image file as another disk in your system."

A block of code is set as follows:

```
html, body, #map {
  height: 100%;
  margin: 0;
  padding: 0
}
```

When we wish to draw your attention to a particular part of a code block, the relevant lines or items are set in bold:

```
[default]
exten => s,1,Dial(Zap/1|30)
exten => s,2,Voicemail(u100)
exten => s,102,Voicemail(b100)
exten => i,1,Voicemail(s0)
```

Any command-line input or output is written as follows:

```
$ mkdir css
$ cd css
```

Bold: Indicates a new term, an important word, or words that you see onscreen. For instance, words in menus or dialog boxes appear in **bold**. Here is an example: "Select **System info** from the **Administration** panel."

> **Tips or important notes**
> Appear like this.

Get in touch

Feedback from our readers is always welcome.

General feedback: If you have questions about any aspect of this book, email us at customercare@ packtpub.com and mention the book title in the subject of your message.

Errata: Although we have taken every care to ensure the accuracy of our content, mistakes do happen. If you have found a mistake in this book, we would be grateful if you would report this to us. Please visit www.packtpub.com/support/errata and fill in the form.

Piracy: If you come across any illegal copies of our works in any form on the internet, we would be grateful if you would provide us with the location address or website name. Please contact us at copyright@packt.com with a link to the material.

If you are interested in becoming an author: If there is a topic that you have expertise in and you are interested in either writing or contributing to a book, please visit authors.packtpub.com.

Share Your Thoughts

Once you've read *Network Science with Python*, we'd love to hear your thoughts! Scan the QR code below to go straight to the Amazon review page for this book and share your feedback.

https://packt.link/r/1-801-07369-4

Your review is important to us and the tech community and will help us make sure we're delivering excellent quality content.

Download a free PDF copy of this book

Thanks for purchasing this book!

Do you like to read on the go but are unable to carry your print books everywhere?

Is your eBook purchase not compatible with the device of your choice?

Don't worry, now with every Packt book you get a DRM-free PDF version of that book at no cost.

Read anywhere, any place, on any device. Search, copy, and paste code from your favorite technical books directly into your application.

The perks don't stop there, you can get exclusive access to discounts, newsletters, and great free content in your inbox daily

Follow these simple steps to get the benefits:

1. Scan the QR code or visit the link below

https://packt.link/free-ebook/9781801073691

2. Submit your proof of purchase
3. That's it! We'll send your free PDF and other benefits to your email directly

Part 1: Getting Started with Natural Language Processing and Networks

You will gain a foundational knowledge on **Natural Language Processing** (NLP), network science, and social network analysis. These chapters are written to give you the required knowledge for working with language and network data.

This section includes the following chapters:

- *Chapter 1, Introducing Natural Language Processing*
- *Chapter 2, Network Analysis*
- *Chapter 3, Useful Python Libraries*

1

Introducing Natural Language Processing

Why in the world would a **network analysis** book start with **Natural Language Processing (NLP)**?! I expect you to be asking yourself that question, and it's a very good question. Here is why: we humans use language and text to describe the world around us. We write about the people we know, the things we do, the places we visit, and so on. Text can be used to reveal relationships that exist. The relationship between things can be shown via network visualization. It can be studied with network analysis.

In short, text can be used to extract interesting relationships, and networks can be used to study those relationships much further. We will use text and NLP to identify relationships and network analysis and visualization to learn more.

NLP is very useful for creating network data, and we can use that network data to learn network analysis. This book is an opportunity to learn a bit about NLP and network analysis, and how they can be used together.

In explaining NLP at a very high level, we will be discussing the following topics:

- What is NLP?
- Why NLP in a network analysis book?
- A very brief history of NLP
- How has NLP helped me?
- Common uses for NLP
- Advanced uses of NLP
- How can a beginner get started with NLP?

Technical requirements

Although there are a few places in this chapter where I show some code, I do not expect you to write the code yet. These examples are only for a demonstration to give a preview of what can be done. The rest of this book will be very hands-on, so take a look and read to understand what I am doing. Don't worry about writing code yet. First, learn the concepts.

What is NLP?

NLP is a set of techniques that helps computers work with human language. However, it can be used for more than dealing with words and sentences. It can also work with application log files, source code, or anything else where human text is used, and on imaginary languages as well, so long as the text is consistent in following a language's rules. *Natural language* is a language that humans speak or write. *Processing* is the act of a computer using data. So, *NLP* is the act of a computer using spoken or written human language. It's that simple.

Many of us software developers have been doing NLP for years, maybe even without realizing it. I will give my own example. I started my career as a web developer. I was entirely self-educated in web development. Early in my career, I built a website that became very popular and had a nice community, so I took inspiration from Yahoo Chats (popular at the time), reverse-engineered it, and built my own internet message board. It grew rapidly, providing years of entertainment and making me some close friends. However, with any good social application, trolls, bots, and generally nasty people eventually became a problem, so I needed a way to flag and quarantine abusive content automatically.

Back then, I created lists of examples of abusive words and strings that could help catch abuse. I was not interested in stopping all obscenities, as I do not believe in completely controlling how people post text online; however, I was looking to identify toxic behavior, violence, and other nasty things. Anyone with a comment section on their website is very likely doing something similar in order to moderate their website, or they should be. The point is that I have been doing NLP since the beginning of my career without even noticing, but it was rule-based.

These days, machine learning dominates the NLP landscape, as we are able to train models to detect abuse, violence, or pretty much anything we can imagine, which is one thing that I love the most about NLP. I feel that I am limited only by the extent of my own creativity. As such, I have created classifiers to detect discussions that contained or were about extreme political sentiment, violence, music, art, data science, natural sciences, and disinformation, and at any given moment, I typically have several NLP models in mind that I want to build but haven't found time. I have even used NLP to detect malware. But, again, NLP doesn't have to be against written or spoken words, as my malware classifier has shown. If you keep that in mind, then your potential uses for NLP massively expand. My rule of thumb *is that if there are sequences in data that can be extracted as words – even if they are not words – they can potentially be used with NLP techniques.*

In the past, and probably still now, analysts would drop columns containing text or do very basic transformations or computations, such as one-hot encoding, counts, or determining the presence/absence (true/false). However, there is so much more that you can do, and I hope this chapter and book will ignite some inspiration and curiosity in you from reading this.

Why NLP in a network analysis book?

Most of you probably bought this book in order to learn applied social network analysis using Python. So, why am I explaining NLP? Here's why: if you know your way around NLP and are comfortable extracting data from text, that can be extremely powerful for creating network data and investigating the relationship between things that are mentioned in text. Here is an example from the book *Alice's Adventures in Wonderland* by Lewis Carroll, my favorite book.

> *"Once upon a time there were three little sisters" the Dormouse began in a great hurry; "and their names were Elsie, Lacie, and Tillie; and they lived at the bottom of a well."*

What can we observe from these words? What characters or places are mentioned? We can see that the Dormouse is telling a story about three sisters named *Elsie*, *Lacie*, and *Tillie* and that they lived at the bottom of a well. If you allow yourself to think in terms of relationships, you will see that these relationships exist:

- Three sisters -> Dormouse (he either knows them or knows a story about them)
- Dormouse -> Elsie
- Dormouse -> Lacie
- Dormouse -> Tillie
- Elsie -> bottom of a well
- Lacie -> bottom of a well
- Tillie -> bottom of a well

It's also very likely that the three sisters all know each other, so additional relationships emerge:

- Elsie -> Lacie
- Elsie -> Tillie
- Lacie -> Elsie
- Lacie -> Tillie
- Tillie -> Elsie
- Tillie -> Lacie

Our minds build these relationship maps so effectively that we don't even realize that we are doing it. The moment I read that the three were sisters, I drew a mental image that the three knew each other.

Let's try another example from a current news story: *Ocasio-Cortez doubles down on Manchin criticism* (CNN, June 2021: `https://edition.cnn.com/videos/politics/2021/06/13/alexandria-ocasio-cortez-joe-manchin-criticism-sot-sotu-vpx.cnn`).

> *Rep. Alexandria Ocasio-Cortez (D-NY) says that Sen. Joe Manchin (D-WV)*
> *not supporting a house voting rights bill is being influenced by the legislation's*
> *sweeping reforms to limit the role of lobbyists and the influence of "dark money"*
> *political donations.*

Who is mentioned, and what is their relationship? What can we learn from this short text?

- Rep. Alexandria Ocasio-Cortez is talking about Sen. Joe Manchin

- Both are Democrats

- Sen. Joe Manchin does not support a house voting rights bill

- Rep. Alexandria Ocasio-Cortez claims that Sen. Joe Manchin is being influenced by the legislation's reforms

- Rep. Alexandria Ocasio-Cortez claims that Sen. Joe Manchin is being influenced by "dark money" political donations

- There may be a relationship between Sen. Joe Manchin and "dark money" political donors

We can see that even a small amount of text has a lot of information embedded.

If you are stuck trying to figure out relationships when dealing with text, I learned in college creative writing classes to consider the "*W*" questions (and *How*) in order to explain things in a story:

- Who: Who is involved? Who is telling the story?

- What: What is being talked about? What is happening?

- When: When does this take place? What time of the day is it?

- Where: Where is this taking place? What location is being described?

- Why: Why is this important?

- How: How is the thing being done?

If you ask these questions, you will notice relationships between things and other things, which is foundational for building and analyzing networks. If you can do this, you can identify relationships in text. If you can identify relationships in text, you can use that knowledge to build social networks. If you can build social networks, you can analyze relationships, detect importance, detect weaknesses, and use this knowledge to gain a really profound understanding of whatever it is that you are analyzing.

You can also use this knowledge to attack dark networks (crime, terrorism, and so on) or protect people, places, and infrastructure. This isn't just insights. *These are actionable insights*—the best kind.

That is the point of this book. Marrying NLP with social network analysis and data science is extremely powerful for acquiring a new perspective. If you can scrape or get the data you need, you can really gain deep knowledge of how things relate and why.

That is why this chapter aims to explain very simply what NLP is, how to use it, and what it can be used for. But before that, let's get into the history for a bit, as that is often left out of NLP books.

A very brief history of NLP

If you research the history of NLP, you will not find one conclusive answer as to its origins. As I was planning the outline for this chapter, I realized that I knew quite a bit about the uses and implementation of NLP but that I had a blind spot regarding its history and origins. I knew that it was tied to computational linguistics, but I did not know the history of that field, either. The earliest conceptualization of **Machine Translation (MT)** supposedly took place in the seventeenth century; however, I am deeply skeptical that this was the origin of the idea of MT or NLP, as I bet people have been puzzling over the relationships between words and characters for as long as language has existed. I would assume that to be unavoidable, as people thousands of years ago were not simpletons. They were every bit as clever and inquisitive as we are, if not more. However, let me give some interesting information I have dug up on the origins of NLP. Please understand that this is not the complete history. An entire book could be written about the origins and history of NLP. So that I quickly move on, I am going to keep this brief. I am going to just list some of the highlights that I found. If you want to know more, this is a rich topic for research.

One thing that puzzles me is that I rarely see cryptology (cryptography and cryptanalysis) mentioned as being part of the origins of NLP or even MT when cryptography is the act of translating a message into gibberish, and cryptanalysis is the act of reversing secret gibberish into a useful message. So, to me, any automation, even hundreds or thousands of years ago, that could assist in carrying out cryptography or cryptanalysis should be part of the conversation. It might not be MT in the same way that modern translation is, but it is a form of translation, nonetheless. So, I would suggest that MT goes back even to the Caesar cipher invented by Julius Caesar, and probably much earlier than that. The Caesar cipher translated a message into code by shifting the text by a certain number. As an example, let's take the sentence:

I really love NLP.

First, we should probably remove the spaces and casing so that any eavesdropper can't get hints on word boundaries. The string is now as follows:

ireallylovenlp

If we do a `shift-1`, we shift each letter by one character to the right, so we get:

jsfbmmzmpwfomq

The number that we shift is arbitrary. We could also use a reverse shift. Wooden sticks were used for converting text into code, so I would consider that as a translation tool.

After the Caesar cipher, many, many other techniques were invented for encrypting human text, some of which were quite sophisticated. There is an outstanding book called *The Code Book* by Simon Singh that goes into the several thousand-year-long history of cryptology. With that said, let's move on to what people typically think of with regard to NLP and MT.

In the seventeenth century, philosophers began to submit proposals for codes that could be used to relate words between languages. This was all theoretical, and none of them were used in the development of an actual machine, but ideas such as MT came about first by considering future possibilities, and then implementation was considered. A few hundred years later, in the early 1900s, Ferdinand de Saussure, a Swiss linguistics professor, developed an approach for describing language as a system. He passed away in the early 1900s and almost deprived the world of the concept of *language as a science*, but realizing the importance of his ideas, two of his colleagues wrote the *Cours de linguistique generale* in 1916. This book laid the foundation for the structuralist approach that started with linguistics but eventually expanded to other fields, including computers.

Finally, in the 1930s, the first patents for MT were applied for.

Later, World War II began, and this is what caused me to consider the Caesar cipher and cryptology as early forms of MT. During World War II, Germany used a machine called the Enigma machine to encrypt German messages. The sophistication of the technique made the codes nearly unbreakable, with devastating effects. In 1939, along with other British cryptanalysts, Alan Turing designed the bombe after the Polish bomba that had been decrypting Enigma messages the seven years prior. Eventually, the bombe was able to reverse German codes, taking away the advantage of secrecy that German U-boats were enjoying and saving many lives. This is a fascinating story in itself, and I encourage readers to learn more about the effort to decrypt messages that were encrypted by the Enigma machines.

After the war, research into MT and NLP really took off. In 1950, Alan Turing published *Computing Machinery and Intelligence*, which proposed the Turing Test as a way of assessing intelligence. To this day, the Turing Test is frequently mentioned as a criterion of intelligence for **Artificial Intelligence (AI)** to be judged by.

In 1954, the Georgetown experiment fully automated translations of more than sixty Russian sentences into English. In 1957, Noam Chomsky's *Syntactic Structures* revolutionized linguistics with a rule-based system of syntactic structures known as **Universal Grammar (UG)**.

To evaluate the progress of MT and NLP research, the US **National Research Council (NRC)** created the **Automatic Language Processing Advisory Committee (ALPAC)** in 1964. At the same time, at MIT, Joseph Weizenbaum had created *ELIZA*, the world's first chatbot. Based on reflection techniques and simple grammar rules, ELIZA was able to rephrase any sentence into another sentence as a response to users.

Then winter struck. In 1966, due to a report by ALPAC, an NLP stoppage occurred, and funding for NLP and MT was discontinued. As a result, AI and NLP research were seen as a dead end by many people, but not all. This freeze lasted until the late 1980s, when a new revolution in NLP would begin, driven by a steady increase in computational power and the shift to **Machine Learning (ML)** algorithms rather than hard-coded rules.

In the 1990s, the popularity of statistical models for NLP arose. Then, in 1997, **Long Short-Term Memory (LSTM)** and **Recurrent Neural Network (RNN)** models were introduced, and they found their niche for voice and text processing in 2007. In 2001, Yoshua Bengio and his team provided the first feed-forward neural language model. In 2011, Apple's Siri became known as one of the world's first successful AI and NLP assistants to be used by general consumers.

Since 2011, NLP research and development has exploded, so this is as far as I will go into history. I am positive that there are many gaps in the history of NLP and MT, so I encourage you to do your own research and really dig into the parts that fascinate you. I have spent much of my career working in cyber security, so I am fascinated by almost anything having to do with the history of cryptology, especially old techniques for cryptography.

How has NLP helped me?

I want to do more than show you how to do something. I want to show you how it can help you. The easiest way for me to explain how this may be useful to you is to explain how it has been useful to me. There are a few things that were really appealing to me about NLP.

Simple text analysis

I am really into reading and grew up loving literature, so when I first learned that NLP techniques could be used for text analysis, I was immediately intrigued. Even something as simple as counting the number of times a specific word is mentioned in a book can be interesting and spark curiosity. For example, how many times is Eve, the first woman mentioned in the Bible, mentioned in the book of Genesis? How many times is she mentioned in the entire Bible? How many times is Adam mentioned in Genesis? How many times is Adam mentioned in the entire Bible? For this example, I'm using the King James Version.

Let's compare:

Name	Genesis Count	Bible Count
Eve	2	4
Adam	17	29

Figure 1.1 – Table of biblical mentions of Adam and Eve

These are interesting results. Even if we do not take the Genesis story as literal truth, it's still an interesting story, and we often hear about Adam and Eve. Hence, it is easy to assume they would be mentioned as frequently, but Adam is actually mentioned over eight times as often as Eve in Genesis and over seven times as often in the entire Bible. Part of understanding literature is building a mental map of what is happening in the text. To me, it's a little odd that Eve is mentioned so rarely, and it makes me want to investigate the amount of male versus female mentions or maybe investigate which books of the Bible have the largest number of female characters and then what those stories are about. If nothing else, it sparks curiosity, which should lead to deeper analysis and understanding.

NLP gives me tools to extract quantifiable data from raw text. It empowers me to use that quantifiable data in research that would have been impossible otherwise. Imagine how long it would have taken to do this manually, reading every single page without missing a detail, to get to these small counts. Now, consider that this took me about a second once the code was written. That is powerful, and I can use this functionality to research any person in any text.

Community sentiment analysis

Second, and related to the point I just made, NLP provides ways to investigate themes and sentiments carried by groups of people. During the Covid-19 pandemic, a group of people has been vocally anti-mask, spreading fear and misinformation. If I capture text from those people, I can use sentiment analysis techniques to determine and measure sentiment shared by that group of people across various topics. I did exactly that. I was able to scrape thousands of tweets and understand what they really felt about various topics such as Covid-19, the flag, the Second Amendment, foreigners, science, and many more. I did this exact analysis for one of my projects, *#100daysofnlp*, on LinkedIn (`https://www.linkedin.com/feed/hashtag/100daysofnlp/`), and the results were illuminating.

NLP allows us to objectively investigate and analyze group sentiment about anything, so long as we are able to acquire text or audio. Much of Twitter data is posted openly and is consumable by the public. Just one caveat: if you are going to scrape, please use your abilities for good, not evil. Use this to understand what people are thinking about and what they feel. Use it for research, not surveillance.

Answer previously unanswerable questions

Really, what ties these two together is that NLP helps me answer questions that were previously unanswerable. In the past, we could have conversations discussing what people felt and why or describing literature we had read but only at a surface level. With what I am going to show you how to do in this book, you will no longer be limited to the surface. You will be able to map out complex relationships that supposedly took place thousands of years ago very quickly, and you will be able to closely analyze any relationship and even the evolution of relationships. You will be able to apply these same techniques to any kind of text, including transcribed audio, books, news articles, and social media posts. NLP opens up a universe of untapped knowledge.

Safety and security

In 2020, the Covid-19 pandemic hit the entire world. When it hit, I was worried that many people would lose their jobs and their homes, and I feared that the world would spin out of control into total anarchy. It's gotten bad but we do not have armed gangs raiding towns and communities around the world. Tension is up, but I wanted a way to keep an eye on violence in my area in real time. So, I scraped police tweets from several police accounts in my area, as they report all kinds of crime in near real time, including violence. I created a dataset of violent versus non-violent tweets, where violent tweets contained words such as shooting, stabbing, and other violence-related words. I then trained an ML classifier to detect tweets having to do with violence. Using the results of this classifier, I can keep an eye on violence in my area. I can keep an eye on anything that I want, so long as I can get text but knowing how my area is doing in terms of street violence could serve as a warning or give me comfort. Again, replacing what was limited to feeling and emotion with quantifiable data is powerful.

Common uses for NLP

One thing that I like the most about NLP is that you are primarily limited by your imagination and what you can do with it. If you are a creative person, you will be able to come up with many ideas that I have not explained.

I will explain some of the common uses of NLP that I have found. Some of this may not typically appear in NLP books, but as a lifelong programmer, when I think of NLP, I automatically think about any programmatic work with string, with a string being a sequence of characters. ABCDEFG is a *string*, for instance. A is a *character*.

> **Note**
> Please don't bother writing the code for now unless you just want to experiment with some of your own data. The code in this chapter is just to show what is possible and what the code may look like. We will go much deeper into actual code throughout this book.

True/False – Presence/Absence

This may not fit strictly into NLP, but it is very often a part of any text operation, and this also happens in ML used in NLP, where one-hot encoding is used. Here, we are looking strictly for the presence or absence of something. For instance, as we saw earlier in the chapter, I wanted to count the number of times that Adam and Eve appeared in the Bible. I could have similarly written some simple code to determine whether Adam and Eve were in the Bible at all or whether they were in the book of Exodus.

For this example, let's use this DataFrame that I have set up:

	book	chapter	verse	text	entities
0	genesis	1	1	In the beginning God created the heaven and th...	God
1	genesis	1	2	And the earth was without form, and void; and ...	Spirit
1	genesis	1	2	And the earth was without form, and void; and ...	God
2	genesis	1	3	And God said, Let there be light: and there wa...	God
3	genesis	1	4	And God saw the light, that it was good: and G...	God

Figure 1.2 – pandas DataFrame containing the entire King James Version text of the Bible

I specifically want to see whether Eve exists as one of the entities in df['entities']. I want to keep the data in a DataFrame, as I have uses for it, so I will just do some pattern matching on the entities field:

```
check_df['entities'].str.contains('^Eve$')
0         False
1         False
1         False
2         False
3         False
          ...
31101     False
31101     False
31102     False
31102     False
31102     False
Name: entities, Length: 51702, dtype: bool
```

Here, I am using what is called a **regular expression (regex)** to look for an exact match on the word Eve. The ^ symbol means that the E in Eve sits at the very beginning of the string, and $ means that the e sits at the very end of the string. This ensures that there is an entity that is exactly named Eve, with nothing before and after. With regex, you have a lot more flexibility than this, but this is a simple example.

In Python, if you have a series of True and False values, .min() will give you False, and .max() will give you True, and that makes sense as another way of looking at True and False is a 1 and 0, and 1 is greater than 0. There are other ways to do this, but I am going to do it this way. So, to see whether Eve is mentioned even once in the whole Bible, I can do the following:

```
check_df['entities'].str.contains('^Eve$').max()
True
```

If I want to see if Adam is in the Bible, I can replace Eve with Adam:

```
check_df['entities'].str.contains('^Adam$').max()
True
```

Detecting the presence or absence of something in a piece of text can be useful. For instance, if we want to very quickly get a list of Bible verses that are about **Eve**, we can do the following:

```
check_df[check_df['entities'].str.contains('^Eve$')]
```

This will give us a DataFrame of Bible verses mentioning Eve:

	book	chapter	verse	text	entities
75	genesis	3	20	And Adam called his wife's name Eve; because s...	Eve
80	genesis	4	1	And Adam knew Eve his wife; and she conceived,...	Eve
28992	2 corinthians	11	3	But I fear, lest by any means, as the serpent ...	Eve
29729	1 timothy	2	13	For Adam was first formed, then Eve.	Eve

Figure 1.3 – Bible verses containing strict mentions of Eve

If we want to get a list of verses that are about Noah, we can do this:

```
check_df[check_df['entities'].str.contains('^Noah$')].head(10)
```

This will give us a DataFrame of Bible verses mentioning **Noah**:

	book	chapter	verse	text	entities
134	genesis	5	29	And he called his name Noah, saying, This same...	Noah
135	genesis	5	30	And Lamech lived after he begat Noah five hund...	Noah
137	genesis	5	32	And Noah was five hundred years old: and Noah ...	Noah
137	genesis	5	32	And Noah was five hundred years old: and Noah ...	Noah
145	genesis	6	8	But Noah found grace in the eyes of the LORD.	Noah
146	genesis	6	9	These are the generations of Noah: Noah was a ...	Noah
146	genesis	6	9	These are the generations of Noah: Noah was a ...	Noah
146	genesis	6	9	These are the generations of Noah: Noah was a ...	Noah
147	genesis	6	10	And Noah begat three sons, Shem, Ham, and Japh...	Noah
150	genesis	6	13	And God said unto Noah, The end of all flesh i...	Noah

Figure 1.4 – Bible verses containing strict mentions of Noah

I have added `.head(10)` to only see the first ten rows. With text, I often find myself wanting to see more than the default five rows.

And if we didn't want to use the **entities** field, we could look in the text instead.

```
df[df['text'].str.contains('Eve')]
```

This will give us a DataFrame of Bible verses where the text of the verse included a mention of **Eve**.

	book	chapter	verse	text	entities
75	genesis	3	20	And Adam called his wife's name Eve; because s...	[Adam, Eve]
80	genesis	4	1	And Adam knew Eve his wife; and she conceived,...	[Adam, Eve, Cain, LORD]
202	genesis	8	19	Every beast, every creeping thing, and every f...	[]
208	genesis	9	3	Every moving thing that liveth shall be meat f...	[]
243	genesis	10	9	He was a mighty hunter before the LORD: wheref...	[LORD, Nimrod, LORD]
...	...	...	...	...	...
30605	1 john	4	2	Hereby know ye the Spirit of God: Every spirit...	[Hereby, Spirit, God, Jesus, Christ, God]
30680	jude	1	7	Even as Sodom and Gomorrha, and the cities abo...	[Sodom, Gomorrha]
30705	revelation	1	7	Behold, he cometh with clouds; and every eye s...	[Behold, Amen]
30962	revelation	16	7	And I heard another out of the altar say, Even...	[Lord, God, Almighty]
31101	revelation	22	20	He which testifieth these things saith, Surely...	[Amen, Lord, Jesus]

Figure 1.5 – Bible verses containing mentions of Eve

That is where this gets a bit messy. I have already done some of the hard work, extracting entities for this dataset, which I will show how to do in a later chapter. When you are dealing with raw text, regex and pattern matching can be a headache, as shown in the preceding figure. I only wanted the verses that contained Eve, but instead, I got matches for words such as **even** and **every**. That's not what I want.

Anyone who works with text data is going to want to learn the basics of regex. Take heart, though. I have been using regex for over twenty years, and I still very frequently have to Google search to get mine working correctly. I'll revisit regex, but I hope that you can see that it is pretty simple to determine if a word exists in a string. For something more practical, if you had 400,000 scraped tweets and you were only interested in the ones that were about a specific thing, you could easily use the preceding techniques or regex to look for an exact or close match.

Regular expressions (regex)

I briefly explained regex in the previous section, but there is much more that you can use it for than to simply determine the presence or absence of something. For instance, you can also use regex to extract data from text to enrich your datasets. Let's look at a data science feed that I scrape:

	created_at	text	publisher
19054	2021-06-16 01:08:00+00:00	Learn about the latest advancements and trends...	odsc
19054	2021-06-16 01:08:00+00:00	Learn about the latest advancements and trends...	odsc
19055	2021-06-16 01:29:59+00:00	How To Become A Data Engineer: From Analyst To...	datasciencedojo
19056	2021-06-16 01:32:01+00:00	One roadblock in using neural networks are the...	tdatascience
19057	2021-06-16 01:37:59+00:00	Discover how Uber scaled their API gateway to ...	odsc

Figure 1.6 – Scraped data science Twitter feed

There's a lot of value in that text field, but it is difficult to work with in its current form. What if I only want a list of links that are posted every day? What if I want to see the hashtags that are used by the data science community? What if I want to take these tweets and build a social network to analyze who interacts? The first thing we should do is enrich the dataset by extracting things that we want. So, if I wanted to create three new fields that contained lists of hashtags, mentions, and URLs, I could do the following:

```
df['text'] = df['text'].str.replace('@', ' @')
df['text'] = df['text'].str.replace('#', ' #')
df['text'] = df['text'].str.replace('http', ' http')
df['users'] = df['text'].apply(lambda tweet: [token for token
in tweet.split() if token.startswith('@')])
df['tags'] = df['text'].apply(lambda tweet: [token for token in
```

```
tweet.split() if token.startswith('#')])
df['urls'] = df['text'].apply(lambda tweet: [token for token in
tweet.split() if token.startswith('http')])
```

In the first three lines, I am adding a space behind each mention, hashtag, and URL just to give a little breathing room for splitting. In the next three lines, I am splitting each tweet by space and then applying rules to identify mentions, hashtags, and URLs. In this case, I don't use fancy logic. Mentions start with @, hashtags start with #, and URLs start with HTTP (to include HTTPS). The result of this code is that I end up with three additional columns, containing lists of users, tags, and URLs.

If I then use explode() on the users, tags, and URLs, I will get a DataFrame where each individual user, tag, and URL has its own row. This is what the DataFrame looks like after explode():

	created_at	text	publisher
19054	2021-06-16 01:08:00+00:00	Learn about the latest advancements and trends...	odsc
19054	2021-06-16 01:08:00+00:00	Learn about the latest advancements and trends...	odsc
19055	2021-06-16 01:29:59+00:00	How To Become A Data Engineer: From Analyst To...	datasciencedojo
19056	2021-06-16 01:32:01+00:00	One roadblock in using neural networks are the...	tdatascience
19057	2021-06-16 01:37:59+00:00	Discover how Uber scaled their API gateway to ...	odsc

Figure 1.7 – Scraped data science Twitter feed, enriched with users, tags, and URLs

I can then use these new columns to get a list of unique hashtags used:

```
sorted(df['tags'].dropna().str.lower().unique())
['#',
 '#,',
 '#1',
 '#1.',
 '#10',
 '#10...',
 '#100daysofcode',
 '#100daysofcodechallenge',
 '#100daysofcoding',
 '#15minutecity',
 '#16ways16days',
 '#1bestseller',
 '#1m_ai_talents_in_10_years!',
 '#1m_ai_talents_in_10_yrs!',
```

```
'#1m_ai_talents_in_10yrs',
'#1maitalentsin10years',
'#1millionaitalentsin10yrs',
'#1newrelease',
'#2'
```

Clearly, the regex used in my data enrichment is not perfect, as punctuation should not be included in hashtags. That's something to fix. Be warned, working with human language is very messy and difficult to get perfect. We just have to be persistent to get exactly what we want.

Let's see what the unique mentions look like. By unique mentions, I mean the deduplicated individual accounts mentioned in tweets:

```
sorted(df['users'].dropna().str.lower().unique())
['@',
 '@027_7',
 '@0dust_himanshu',
 '@0x72657562656e',
 '@16yashpatel',
 '@18f',
 '@1ethanhansen',
 '@1littlecoder',
 '@1njection',
 '@1wojciechnowak',
 '@20,',
 '@26th_february_2021',
 '@29mukesh89',
 '@2net_software',
```

That looks a lot better, though @ should not exist alone, the fourth one looks suspicious, and a few of these look like they were mistakenly used as mentions when they should have been used as hashtags. That's a problem with the tweet text, not the regular expression, most likely, but worth investigating.

I like to lowercase mentions and hashtags so that it is easier to find unique tags. This is often done as *preprocessing for NLP*.

Finally, let's get a list of unique URLs mentioned (which can then be used for further scraping):

```
sorted(df['urls'].dropna().unique())
['http://t.co/DplZsLjTr4',
 'http://t.co/fYzSPkY7Qk',
```

```
'http://t.co/uDclS4EI98',
'https://t.co/01IIAL6hut',
'https://t.co/01OwdBe4ym',
'https://t.co/01wDUOpeaH',
'https://t.co/026c3qjvcD',
'https://t.co/02HxdLHPSB',
'https://t.co/02egVns8MC',
'https://t.co/02tIoF63HK',
'https://t.co/030eotd619',
'https://t.co/033LGtQMfF',
'https://t.co/034W5ItqdM',
'https://t.co/037UMOuInk',
'https://t.co/039nGOjyZr'
```

This looks very clean. How many URLs was I able to extract?

```
len(sorted(df['urls'].dropna().unique()))
19790
```

That's a lot of links. As this is Twitter data, a lot of URLs are often photos, selfies, YouTube links, and other things that may not be too useful to a researcher, but this is my scraped data science feed, which pulls information from dozens of data science related accounts, so many of these URLs likely include exciting news and research.

Regex allows you to extract additional data from your data and use it to enrich your datasets to do easier or further analysis, and if you extract URLs, you can use that as input for additional scraping.

I'm not going to give a long lesson into regex. There are a whole lot of books dedicated to the topic. It is likely that, eventually, you will need to learn how to use regex. For what we are doing in this book, the preceding regex is probably all you will need, as we are using these tools to build social networks that we can analyze. This book isn't primarily about NLP. We just use some NLP techniques to create or enrich our data, and then we will use network analysis for everything else.

Word counts

Word counts are also useful, especially when we want to compare things against each other. For instance, we already compared the number of times that Adam and Eve were mentioned in the Bible, but what if we want to see the number of times that all entities are mentioned in the Bible? We can do this the simple way, and we can do this the NLP way. I prefer to do things the simple way, where possible, but frequently, the NLP or graph way ends up being the simpler way, so learn everything that you can and decide for yourself.

We will do this the simple way by counting the number of times entities were mentioned. Let's use the dataset again and just do some aggregation to see who the most mentioned people are in the Bible. Keep in mind we can do this for any feed that we scrape, so long as we have enriched the dataset to contain a list of mentions. But for this demonstration, I'll use the Bible.

On the third line, I am keeping entities with a name longer than two characters, effectively dropping some junk entities that ended up in the data. I am using this as a filter:

```
check_df = df.explode('entities')
check_df.dropna(inplace=True) # dropping nulls
check_df = check_df[check_df['entities'].apply(len) > 2] #
dropping some trash that snuck in
check_df['entities'] = check_df['entities'].str.lower()
agg_df = check_df[['entities', 'text']].groupby('entities').
count()
agg_df.columns = ['count']
agg_df.sort_values('count', ascending=False, inplace=True)
agg_df.head(10)
```

This is shown in the following DataFrame:

entities	count
lord	7538
god	4354
israel	2564
david	1011
jesus	980
jerusalem	810
judah	808
moses	649
egypt	602
thou	586

Figure 1.8 – Entity counts across the entire Bible

This looks pretty good. Entities are people, places, and things, and the only oddball in this bunch is the word **thou**. The reason that snuck in is that in the Bible, often the word *thou* is capitalized as *Thou*, which gets tagged as an **NNP** (**Proper Noun**) when doing entity recognition and extraction in NLP. However, *thou* is in reference to *You*, and so it makes sense. For example, *Thou shalt not kill, thou shalt not steal.*

If we have the data like this, we can also very easily visualize it for perspective:

```
agg_df.head(10).plot.barh(figsize=(12, 6), title='Entity Counts
by Bible Mentions', legend=False).invert_yaxis()
```

This will give us a horizontal bar chart of entity counts:

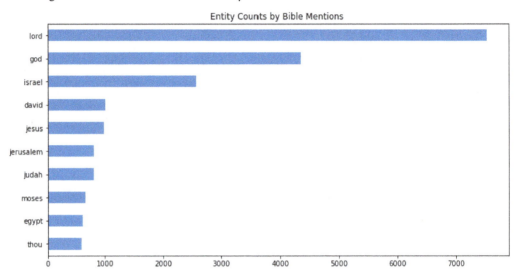

Figure 1.9 – Visualized entity counts across the entire Bible

This is obviously not limited to use on the Bible. If you have any text at all that you are interested in, you can use these techniques to build a deeper understanding. If you want to use these techniques to pursue art, you can. If you want to use these techniques to help fight crime, you can.

Sentiment analysis

This is my favorite technique in all of NLP. I want to know what people are talking about and how they feel about it. This is an often underexplained area of NLP, and if you pay attention to how most people use it, you will see many demonstrations on how to build classifiers that can determine positive or negative sentiment. However, we humans are complicated. We are not just happy or sad. Sometimes, we are neutral. Sometimes we are mostly neutral but more positive than negative. Our feelings are nuanced. One book that I have used a lot for my own education and research into sentiment analysis mentions a study that mapped out human emotions as having primary, secondary, and tertiary emotions (Liu, *Sentiment Analysis*, 2015, p. 35). Here are a few examples:

Primary Emotion	Secondary Emotion	Tertiary Emotion
Anger	Disgust	Contempt
Anger	Envy	Jealousy
Fear	Horror	Alarm
Fear	Nervousness	Anxiety
Love	Affection	Adoration
Love	Lust	Desire

Figure 1.10 – A table of primary, secondary, and tertiary emotions

There are a few primary emotions, there are more secondary emotions, and there are many, many more tertiary emotions. Sentiment analysis can be used to try to classify yes/no for any emotion as long as you have training data.

Sentiment analysis doesn't have to only be used for detecting emotions. The techniques can also be used for classification, so I don't feel that sentiment analysis is quite the complete wording, and maybe that is why there are so many demonstrations of people simply detecting positive/negative sentiment from Yelp and Amazon reviews.

I have more interesting uses for sentiment classification. Right now, I use these techniques to detect toxic speech (really abusive language), positive sentiment, negative sentiment, violence, good news, bad news, questions, disinformation research, and network science research. You can use this as intelligent pattern matching, which learns the nuances of how text about a topic is often written about. For instance, if we wanted to catch tweets related to disinformation, we could train a model on text having to do with misinformation, disinformation, and fake news. The model would learn other related terms during training, and it would do a much better and much faster job of catching them than any human could.

Sentiment analysis and text classification advice

Here is some advice before I move on to the next section: for sentiment analysis and text classification, in many cases, you do not need a neural network for something this simple. If you are building a classifier to detect hate speech, a "bag of words" approach will work for preprocessing, and a simple model will work for classification. Always start simple. A neural network may give you a couple of percents better accuracy if you work at it, but it'll take more time and be less explainable. A `linearsvc` model can be trained in a split second and often do as well, sometimes even better, and some other simple models and techniques should be attempted as well.

Another piece of advice: experiment with stopword removal, but don't just remove stopwords because that's what you have been told. Sometimes it helps, and sometimes it hurts your model. The majority of the time, it might help, but it's simple enough to experiment.

Also, when building your datasets, you can often get the best results if you do sentiment analysis against sentences rather than large chunks of text. Imagine that we have the following text:

Today, I woke up early, had some coffee, and then I went outside to check on the flowers. The sky was blue, and it was a nice, warm June morning. However, when I got back into the house, I found that a pipe had sprung a leak and flooded the entire kitchen. The rest of the day was garbage. I am so angry right now.

Do you think that the emotions of the first sentence are identical to the emotions of the last sentence? This imaginary story is all over the place, starting very cheerful and positive and ending in disaster and anger. If you classify at the sentence level, you are able to be more precise. However, even this is not perfect.

Today started out perfectly, but everything went to hell and I am so angry right now.

What is the sentiment of that sentence? Is it positive or negative? It's both. And so, ideally, if you could capture that a sentence has multiple emotions on display, that would be powerful.

Finally, when you build your models, you always have the choice of whether you want to build binary or multiclass language models. For my own uses, and according to research that has resonated with me, it is often easiest to build small models that simply look for the presence of something. So, rather than building a neural network to determine whether the text is positive, negative, or neutral, you could build a model that looks for positive versus other and another one that looks for negative versus other.

This may seem like more work, but I find that it goes much faster, can be done with very simple models, and the models can be chained together to look for an array of different things. For instance, if I wanted to classify political extremism, I could use three models: toxic language, politics, and violence. If a piece of text was classified as positive for toxic language, was political, and was advocating violence, then it is likely that the poster may be showing some dangerous traits. If only toxic language and political sentiment were being displayed, well, that's common and not usually politically extreme or dangerous. Political discussion is often hostile.

Information extraction

We have already done some information extraction in previous examples, so I will keep this brief. In the previous section, we extracted user mentions, hashtags, and URLs. This was done to enrich the dataset, making further analysis much easier. I added the steps that extract this into my scrapers directly so that I have the lists of users, mentions, and URLs immediately when I download fresh data. This allows me to immediately jump into network analysis or investigate the latest URLs. Basically, if there is information you are looking for, and you come up with a way to repeatedly extract it from text, and you find yourself repeating the steps over and over on different datasets, you should consider adding that functionality to your scrapers.

The most powerful data that is enriching my Twitter datasets are two fields: *publisher*, and *users*. **Publisher** is the account that posted the tweet. **Users** are the accounts mentioned by the publisher. Each of my feeds has dozens of publishers. With publishers and users, I can build social networks from raw text, which will be explained in this book. It is one of the most useful things I have figured out how to do, and you can use the results to find other interesting accounts to scrape.

Community detection

Community detection is not typically mentioned with regard to NLP, but I do think that it should be, especially when using social media text. For instance, if we know that certain hashtags are used by certain groups of people, we can detect other people that may be affiliated with or are supporters of those groups by the hashtags that they use. It is very easy to use this to your advantage when researching groups of people. Just scrape a bunch of them, see what hashtags they use, then search those hashtags. Mentions can give you hints of other accounts to scrape as well.

Community detection is commonly mentioned in social network analysis, but it can also very easily be done with NLP, and I have used topic modeling and the preceding approach as ways of doing so.

Clustering

Clustering is a technique commonly found in unsupervised learning but also done in network analysis. In clustering, we are looking for things that are similar to other things. There are different approaches for doing this, and even NLP topic modeling can be used as clustering. In unsupervised ML, you can use algorithms such as k-means to find tweets, sentences, or books that are similar to other tweets, sentences, or books. You could do similar with topic modeling, using TruncatedSVD. Or, if you have an actual sociogram (map of a social network), you could look at the connected components to see which nodes are connected or apply k-means against certain network metrics (we will go into this later) to see nodes with similar characteristics.

Advanced uses of NLP

Most of the NLP that you will do on a day-to-day basis will probably fall into one of the simpler uses, but let's also discuss a few advanced uses. In some cases, what I am describing as advanced uses is really a combination of simpler uses to deliver a more complicated solution. So, let's discuss some of the more advanced uses, such as chatbots and conversational agents, language modeling, information retrieval, text summarization, topic discovery and modeling, text-to-speech and speech-to-text conversion, MT, and personal assistants.

Chatbots and conversational agents

A **chatbot** is a program that can hold a conversation with its user. These have existed for years, with the earliest chatbots being created in the 1960s, but they have been steadily improving and are now an effective means of forwarding a user to a more specific form of customer support, for instance. If you go to a website's support section, you may be presented with a small chatbox popup that'll say something like, "*What can we help you with today?*". You may then type, "*I want to pay off my credit card balance.*" When the application receives your answer, it'll be able to use it to determine the correct form of support that you need.

While a chatbot is built to handle human text, a **conversational agent** can handle voice audio. Siri and Alexa are examples of conversational agents. You can talk to them and ask them questions.

Chatbots and conversational agents are not limited to text, though; we often run into similar switchboards when we call a company with our phones. We will get the same set of questions, either looking for a word answer or numeric input. So, behind the scenes, where voice is involved, there will be a voice-to-text conversion element in place. Applications also need to determine whether someone is asking a question or making a statement, so there is likely text classification involved as well.

Finally, to provide an answer, text summarization could convert search results into a concise statement to return as text or speech to the user to complete the interaction.

Chatbots are not just rudimentary question-and-answer systems, though. I believe that they will be an effective way for us to interact with text. For instance, you could build a chatbot around the book *Alice's Adventures in Wonderland* (or the Bible) to give answers to questions specifically about the book. You could build a chatbot from your own private messages and talk to yourself. There's a lot of room for creativity here.

Language modeling

Language modeling is concerned with predicting the next word given a sequence of words. For instance, what would come after this: "*The cow jumped over the* _____." Or this: "*Green eggs and* _____." If you go to Google and start typing in the search bar, take note that the next predicted word will show in a drop-down list to speed up your search.

We can see that Google has predicted the next word to be **ham** but that it is also finding queries that are related to what you have already typed. This looks like a combination of language modeling as well as clustering or topic modeling. They are looking to predict the next word before you even type it, and they are even going a step further to find other queries that are similar to the text you have already typed.

Data scientists are also able to use language modeling as part of the creation of generative models. In 2020, I trained a model on thousands of lines of many Christmas songs and trained it to write Christmas poetry. The results were rough and humorous, as I only spent a couple of days on it, but it was able to take seed text and use it to create entire poems. For instance, seed text could be, "Jingle bells," and then the model would continuously take previous text and use it to create a poem until it reached the end limits for words and lines. Here is my favorite of the whole batch:

```
youre the angel
and the manger is the king
```

```
of kings and sing the sea
of door to see to be
the christmas night is the world
and i know i dont want
to see the world and much
see the world is the sky
of the day of the world
and the christmas is born and
the world is born and the
world is born to be the
christmas night is the world is
born and the world is born
to you to you to you
to you to you to a
little child is born to be
the world and the christmas tree
is born in the manger and
everybody sing fa
la la la la la la
la la la la la la
la la la la la la
la la la la la la
la la la la la la
```

I built the generative process to take a random first word from any of the lines of the training data songs. From there, it would create lines of 6 words for a total of 25 lines. I only trained it for 24 hours, as I wanted to quickly get this done in time for Christmas. There are several books on creating generative models, so if you would like to use AI to augment your own creativity, then I highly suggest looking into them. It feels like a collaboration with a model rather than replacing ourselves with a model.

These days, generative text models are becoming quite impressive. ChatGPT – released in November 2022 – has grabbed the attention of so many people with its ability to answer most questions and give realistic-looking answers. The answers are not always correct, so generative models still have a long way to go, but there is now a lot of hype around generative models, and people are considering how they can use them in their own work and lives and what they mean for our future.

Text summarization

Text summarization is pretty much self-explanatory. The goal is to take text as input and return a summary as output. This can be very powerful when you are managing thousands or millions of documents and want to provide a concise sentence about what is in each. It is essentially returning similar to an "abstract" section you would find in an academic article. Many of the details are removed, and only the essence is returned.

However, this is not a perfect art, so be aware that if you use this, the algorithm may throw away important concepts from the text while keeping those that are of less importance. ML is not perfect, so keep an eye on the results.

However, this is not for search so much as it is for returning a short summary. You can use topic modeling and classification to determine document similarity and then use this to summarize the final set of documents.

If you were to take the content of this entire book and feed it to a text summarization algorithm, I would hope that it would capture that the marriage of NLP and network analysis is powerful and approachable by everyone. You do not need to be a genius to work with ML, NLP, or social network analysis. I hope this book will spark your creativity and make you more effective at problem solving and thinking critically about things. There are a lot of important details in this text, but that is the essence.

Topic discovery and modeling

Topic discovery and modeling is very similar to clustering. This is used in **Latent Semantic Indexing (LSI)**, which can be powerful for identifying themes (topics) that exist in text, and it can also be an effective preprocessing step for text classification, allowing models to be trained on context rather than words alone. I mentioned previously, in the *Clustering* and *Community detection* subsections, how this could be used to identify subtle groups within communities based on the words and hashtags they place into their account descriptions.

For instance, topic modeling would find similar strings in topics. If you were to do topic modeling on political news and social media posts, you will notice that in topics, like attracts like. Words will be found with other similar words. For instance, *2A* may be spelled out as the *Second Amendment, USA* may be written in its expanded form (*United States of America*), and so on.

Text-to-speech and speech-to-text conversion

This type of NLP model aims to convert text into speech audio or audio into text transcripts. This is then used as input into classification or conversational agents (chatbots, personal assistants).

What I mean by that is that you can't just feed audio to a text classifier. Also, it's difficult to capture context from audio alone without any language analysis components, as people speak in different dialects, in different tones, and so on.

The first step is often transcribing the audio into text and then analyzing the text itself.

MT

Judging by the history of NLP, I think it is safe to say that translating from language A to language B has probably been on the minds of humans for as long as we have had to interact with other humans who use a different language. For instance, there is even a story in the Bible about the Tower of Babel and we lost the ability to understand each other's words when it was destroyed. MT has so many useful implications, for collaboration, security, and even creativity.

For instance, for collaboration, you need to be able to share knowledge, even if team members do not share the same language. In fact, this is useful anywhere that sharing knowledge is desirable, so you will often find a **see translation** link in social media posts and comments. Today, MT seems almost perfect, though there are occasionally entertaining mistakes.

For security, you want to know what your enemies are planning. Spying is probably not very useful if you can't actually understand what your enemies are saying or planning on doing. Translation is a specialized skill and is a long and manual process when humans are involved. MT can greatly speed up analysis, as the other language can be translated into your own.

And for creativity, how fun would it be to convert text from one language into your own created language? This is completely doable.

Due to the importance of MT and text generation, massive neural networks have been trained to handle text generation and MT.

Personal assistants

Most of us are probably aware of personal assistants such as Alexa and Siri, as they have become an integral part of our lives. I suspect we are going to become even more dependent on them, and we will eventually talk to our cars like on the old TV show *Knight Rider* (broadcast on TV from 1982 to 1986). "*Hey car, drive me to the grocery store*" will probably be as common as "*Hey Alexa, what's the weather going to be like tomorrow?*"

Personal assistants use a combination of several NLP techniques previously mentioned. They may use classification to determine whether your query is a question or a statement. They may then search the internet to find web content most related to the question that you asked. It can then capture the raw text from one or more of the results and then use summarization to build a concise answer. Finally, it will convert text to audio and speak the answer back to the user.

Personal assistants use a combination of several NLP techniques mentioned previously:

1. They may use classification to determine whether your query is a question or a statement.
2. They may then search the internet to find web content that is most related to the question that you asked.

3. They can then capture the raw text from one or more of the results and then use summarization to build a concise answer.

4. Finally, they will convert text to audio and speak the answer back to the user.

I am very excited about the future of personal assistants. I would love to have a robot and car that I can talk to. I think that creativity is probably our only limitation for the different types of personal assistants that we can create or for the modules that they use.

How can a beginner get started with NLP?

This book will be of little use if we do not eventually jump into how to use these tools and technologies. The common and advanced uses that I described here are just some of the uses. As you become comfortable with NLP, I want you to constantly consider other uses for NLP that are possibly not being met. For instance, in text classification alone, you can go very deep. You could use text classification to attempt to classify even more difficult concepts, such as sarcasm or empathy, for instance, but let's not get ahead of ourselves. This is what I want you to do.

Start with a simple idea

Think simply, and only add complexity as needed. Think of something that interests you that you would like to know more about, and then find people who talk about it. If you are interested in photography, find a few Twitter accounts that talk about it. If you are looking to analyze political extremism, find a few Twitter accounts that proudly show their unifying hashtags. If you are interested in peanut allergy research, find a few Twitter accounts of researchers that post their results and articles in their quest to save lives. I mention Twitter over and over because it is a goldmine for investigating how groups of people talk about issues, and people often post links, which can lead to even more scraping. But you could use any social media platform, as long as you can scrape it.

However, start with a very simple idea. What would you like to know about a piece of text (or a lot of Tweets)? What would you like to know about a community of people? Brainstorm. Get a notebook and start writing down every question that comes to mind. Prioritize them. Then you will have a list of questions to seek answers for.

For instance, my research question could be, "*What are people saying about Black Lives Matter protests?*" Or, we could research something less serious and ask, "*What are people saying about the latest Marvel movie?*" Personally, I prefer to at least attempt to use data science for good, to make the world a bit safer, so I am not very interested in movie reviews, but others are. We all have our preferences. Study what interests you.

For this demonstration, I will use my scraped data science feed. I have a few starter questions:

*   Which accounts post the most frequently every week?

*   Which accounts are mentioned the most?

- Which are the primary hashtags used by this community of people?

- What follow-up questions can we think of after answering these questions?

We will only use NLP and simple string operations to answer these questions, as I have not yet begun to explain social network analysis. I am also going to assume that you know your way around Python programming and are familiar with the pandas library. I will cover pandas in more detail in a later chapter, but I will not be giving in-depth training. There are a few great books that cover pandas in depth.

Here is what the raw data for my scraped data science feed looks like:

	date	tweet	users	tags	urls	created_week	publisher
5678	2021-03-10 03:23:00+00:00	Getting Started with SageMaker for Model Train...	[]	[]	['https://t.co/GKuQyuGYSZ']	202110	tdatascience
5679	2021-03-10 04:26:01+00:00	Introduction to Credit Risk Modeling by Juhi R...	[]	[]	['https://t.co/IlyVYV5XGj']	202110	tdatascience
5680	2021-03-10 05:16:01+00:00	Human Priors in Object Detection by Javier htt...	[]	[]	['https://t.co/HNU7aOcjga']	202110	tdatascience
5682	2021-03-10 06:31:01+00:00	Box-Cox transformation is the magic we need by...	[]	[]	['https://t.co/eYfDpcFLYB']	202110	tdatascience
5686	2021-03-10 07:54:00+00:00	fit() vs predict() vs fit_predict() in Python ...	['@gmyrianthous']	[]	['https://t.co/L4rBiPSOiV']	202110	tdatascience

Figure 1.11 – Scraped and enriched data science Twitter feed

To save time, I have set up the regex steps in the scraper to create columns for users, tags, and URLs. All of this is scraped or generated as a step during automated scraping. This will make it much easier and faster to answer the four questions I posed. So, let's get to it.

Accounts that post most frequently

The first thing I want to do is see which accounts post the most in total. I will also take a glimpse at which accounts post the least to see whether any of the accounts have dried up since adding them to my scrapers. For this, I will simply take the columns for `publisher` (the account that posted the tweet) and `tweet`, do a `groupby` operation on the publisher, and then take the count:

```
Check_df = df[['publisher', 'tweet']]
check_df = check_df.groupby('publisher').count()
check_df.sort_values('tweet', ascending=False, inplace=True)
check_df.columns = ['count']
check_df.head(10)
```

This will display a DataFrame of publishers by tweet count, showing us the most active publishers:

publisher	count
datasciencedojo	4830
analyticbridge	4073
tdatascience	3673
mitpress	2877
usgs	2540
packtpub	2349
wids_worldwide	2166
datasciencefest	2108
makcocis	2065
odsc	1859

Figure 1.12 – User tweet counts from the data science Twitter feed

That's awesome. So, if you want to break into data science and you use Twitter, then you should probably follow these accounts.

However, to me, this is of limited use. I really want to see each account's posting behavior. For this, I will use a pivot table. I will use `publisher` as the index, `created_week` as the columns, and run a count aggregation. Here is what the top ten looks like, sorted by the current week:

```
Check_df = df[['publisher', 'created_week', 'tweet']].copy()
pvt_df = pd.pivot_table(check_df, index='publisher',
columns='created_week', aggfunc='count').fillna(0)
pvt_df = pvt_df['tweet']
pvt_df.sort_values(202129, ascending=False, inplace=True)
keep_weeks = pvt_df.columns[-13:-1] # keep the last twelve
weeks, but excluding current
pvt_df = pvt_df[keep_weeks]
pvt_df.head(10)
```

This creates the following DataFrame:

created_week	202118	202119	202120	202121	202122	202123	202124	202125	202126	202127	202128	202129
publisher												
odsc	0.0	0.0	0.0	0.0	0.0	0.0	0.0	0.0	312.0	461.0	418.0	496.0
analyticbridge	0.0	0.0	144.0	334.0	0.0	204.0	334.0	242.0	302.0	6.0	336.0	298.0
datasciencedojo	266.0	246.0	261.0	227.0	241.0	272.0	269.0	225.0	271.0	226.0	232.0	216.0
packtpub	106.0	93.0	89.0	153.0	148.0	168.0	111.0	104.0	168.0	208.0	169.0	195.0
tdatascience	180.0	183.0	181.0	180.0	181.0	181.0	189.0	187.0	184.0	179.0	182.0	184.0
usgs	124.0	117.0	145.0	72.0	91.0	99.0	101.0	107.0	86.0	86.0	105.0	79.0
mitpress	145.0	99.0	100.0	76.0	44.0	71.0	90.0	131.0	118.0	103.0	152.0	78.0
lawrennd	2.0	98.0	41.0	12.0	18.0	27.0	119.0	246.0	2.0	95.0	31.0	73.0
tdietterich	83.0	49.0	78.0	55.0	95.0	41.0	62.0	53.0	23.0	6.0	46.0	72.0
womeninstat	127.0	102.0	48.0	55.0	0.0	29.0	24.0	125.0	0.0	51.0	0.0	52.0

Figure 1.13 – Pivot table of user tweet counts by week

This looks much more useful, and it is sensitive to the week. This should also be interesting to see as a visualization, to get a feel for the scale:

```
_= pvt_df.plot.bar(figsize=(13,6), title='Twitter Data Science
Accounts - Posts Per Week', legend=False)
```

We get the following plot:

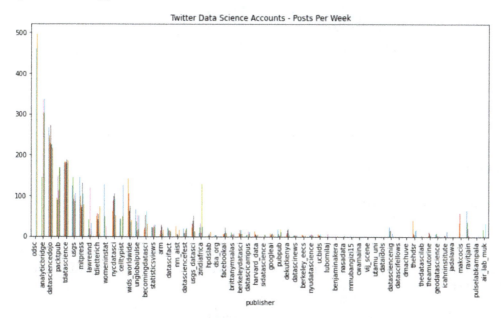

Figure 1.14 – Bar chart of user tweet counts by week

It's a bit difficult to see individual weeks when visualized like this. With any visualization, you will want to think about how you can most easily tell the story that you want to tell. As I am mostly interested in visualizing which accounts post the most in total, I will use the results from the first aggregation instead. This is interesting and cool to look at, but it's not very useful:

```
_= check_df.plot.bar(figsize=(13,6), title='Twitter Data
Science Accounts - Posts Per Week', legend=False)
```

This code gives us the following graph:

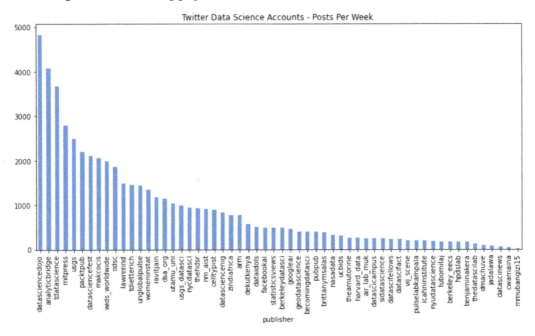

Figure 1.15 – A bar chart of user tweet counts in total

That is much easier to understand.

Accounts mentioned most frequently

Now, I want to see which accounts are mentioned by publishers (the account making the tweet) the most often. This can show people who collaborate, and it can also show other interesting accounts that are worth scraping. For this, I'm just going to use `value_counts` of the top 20 accounts. I want a fast answer:

```
Check_df = df[['users']].copy().dropna()
check_df['users'] = check_df['users'].str.lower()
```

```
check_df.value_counts()[0:20]
users
@dataidols          623
@royalmail          475
@air_lab_muk        231
@makcocis           212
@carbon3it          181
@dictsmakerere      171
@lubomilaj          167
@brittanymsalas     164
@makererenews       158
@vij_scene          151
@nm_aist            145
@makerereu          135
@packtpub           135
@miic_ug            131
@arm                127
@pubpub             124
@deliprao           122
@ucberkeley         115
@mitpress           114
@roydanroy          112
dtype: int64
```

This looks great. I bet there are some interesting data scientists in this bunch of accounts. I should look into that and consider scraping them and adding them to my data science feed.

Top 10 data science hashtags

Next, I want to see which hashtags are used the most often. The code is going to be very similar, other than I need to run `explode()` against the tags field in order to create one row for every element of each tweet's list of hashtags. Let's do that first. For this, we can simply create the DataFrame, drop nulls, lowercase the tags for uniformity, and then use `value_counts()` to get what we want:

```
Check_df = df[['tags']].copy().dropna()
check_df['tags'] = check_df['tags'].str.lower()
check_df.value_counts()[0:10]
tags
```

```
#datascience           2421
#dsc_article           1597
#machinelearning        929
#ai                     761
#wids2021               646
#python                 448
#dsfthegreatindoors     404
#covid19                395
#dsc_techtarget         340
#datsciafrica           308
dtype: int64
```

This looks great. I'm going to visualize the top ten results. However, value_counts() was somehow causing the hashtags to get butchered a bit, so I did a groupby operation against the DataFrame instead:

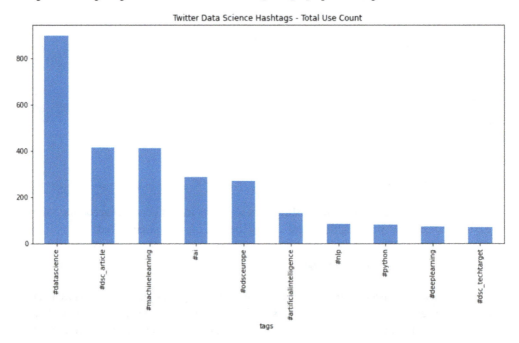

Figure 1.16 – Hashtag counts from the data science Twitter feed

Let's finish up this section with a few more related ideas.

Additional questions or action items from simple analysis

In total, this analysis would have taken me about 10 minutes to do if I was not writing a book. The code might seem strange, as you can chain commands together in Python. I prefer to have significant operations on their own line so that the next person who will have to manage my code will not miss something important that was tacked on to the end of a line. However, notebooks are pretty personal, and notebook code is not typically written with perfectly clean code. When investigating data or doing rough visualizations, focus on what you are trying to do. You do not need to write perfect code until you are ready to write the production version. That said, do not throw notebook quality code into production.

Now that we have done the quick analysis, I have some follow-up questions that I should look into answering:

- How many of these accounts are actually data science related and that I am not already scraping?

- Do any of these accounts give me ideas for new feeds? For instance, I have feeds for data science, disinformation research, art, natural sciences, news, political news, politicians, and more. Maybe I should have a photography feed, for instance.

- Would it be worth scraping by keyword for any of the top keywords to harvest more interesting content and accounts?

- Have any of the accounts dried up (no new posts ever)? Which ones? When did they dry up? Why did they dry up?

You try. Do you have any questions you can think of, given this dataset?

Next, let's try something similar but slightly different, using NLP tools against the book *Alice's Adventures in Wonderland*. Specifically, I want to see whether I can take the `tf-idf` vectors and plot out character appearance by chapter. If you are unfamiliar with it, **term frequency-inverse of document frequency** (TF-IDF) is an appropriate name because that is exactly the math. I won't go into the code, but this is what the results look like:

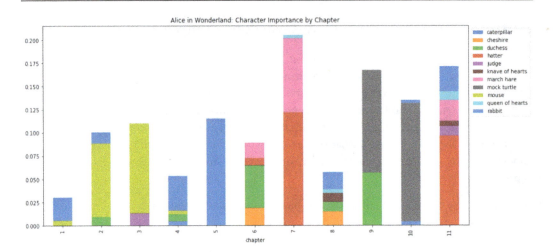

Figure 1.17 – TF-IDF character visualization of Alice's Adventures in Wonderland by book chapter

By using a stacked bar chart, I can see which characters appear together in the same chapters, as well as their relative importance based on the frequency with that they were named. This is completely automated, and I think it would allow for some very interesting applications, such as a more interactive way of researching various books. In the next chapter, I will introduce social network analysis, and if you were to add that in as well, you could even build the social network of Alice in Wonderland, or any other piece of literature, allowing you to see which characters interact.

In order to perform a `tf-idf` vectorization, you need to split sentences apart into tokens. Tokenization is NLP 101 stuff, with a token being a word. So, for instance, if we were to tokenize this sentence:

Today was a wonderful day.

I would end up with a list of the following tokens:

```
['Today', 'was', 'a', 'wonderful', 'day', '.']
```

If you have a collection of several sentences, you can then feed it to `tf-idf` to return the relative importance of each token in a corpus of text. This is often very useful for text classification using simpler models, and it can also be used as input for topic modeling or clustering. However, I have never seen anyone else use it to determine character importance by book chapters, so that's a creative approach.

This example only scratches the surface of what we can do with NLP and investigates only a few of the questions we could come up with. As you do your own research, I encourage you to keep a paper notebook handy, so that you can write down questions to investigate whenever they come to mind.

Summary

In this chapter, we covered what NLP is, how it has helped me, some common and advanced uses of NLP, and how a beginner can get started.

I hope this chapter gave you a rough idea of what NLP is, what it can be used for, what text analysis looks like, and some resources for more learning. This chapter is in no way the complete picture of NLP. It was difficult to even write the history, as there is just so much to the story and so much that has been forgotten over time.

Thank you for reading. This is my first time writing a book, and this is my first chapter ever written for a book, so this really means a lot to me! I hope you are enjoying this so far, and I hope I gave you an adequate first look into NLP.

Next up: network science and social network analysis!

2
Network Analysis

In this chapter, I am going to describe three different topics, but at a very high level: *graph theory*, *social network analysis*, and *network science*. We will begin by discussing some of the confusion around the word *network* and why that will probably remain confusing. Then, we will go over a tiny bit of the past and present of all three. Finally, we will dive into how network analysis has helped me and hopefully how it can help you. This isn't a code-heavy chapter. This is a high-level introduction.

The following topics will be discussed in this chapter:

- The confusion behind networks
- What is this network stuff?
- Resources for learning network analysis
- Common network use cases
- Advanced network use cases
- Getting started with networks

The confusion behind networks

First, to reduce confusion, if you see me mention *NetworkX*, that is not a typo. That is a Python library we will be making heavy use of in this book. Python is a very popular programming language.

I have worked in **information technology** (**IT**) my entire career, and have even gone farther than that. During some points of my career, I pursued security certifications such as Security+ and CISSP for job requirements, and I have constantly worked with other IT professionals, such as network engineers. So, believe me when I tell you that I understand the awkwardness that is involved in discussing network science with people who view networks as primarily based on TCP/IP and subnets.

Networks are all around us and even inside of us. In fact, our brain is the most complex thing we have found in the universe, as discussed in the book *How to Create a Mind* (Kurzweil, 2012). Our brain comprises of hundreds of billions of cells interlinked through trillions of connections. What a complex network!

When I think of all that we have achieved with artificial neural networks and how well they do for translation, computer vision, and generation, I think they are impressive, but I am much more impressed with how these ultra-complex minds of ours naturally evolved, and how foolish we sometimes are even with all of that complexity.

Back to my point, *networks* are not just computer networks. Networks are comprised of relationships and information exchanges between things.

In this book, when I talk about networks, I am talking about *graphs*. I am talking about the complex relationships between things. One of my favorite books on this subject, titled *Networks* (Newman, 2018), describes a few different kinds of networks, such as the following:

- Technological networks

  - Internet

  - Power grid

  - Telephone networks

  - Transportation networks

    - Airline routes

    - Roads

    - Rail networks

  - Delivery and distribution networks

- Information networks

  - World wide web

  - Citation networks

  - Peer-to-peer networks

  - Recommender networks

  - Keyword indexes

  - Dataflow networks

- Social networks

  - Social interaction networks

  - Ego-centric networks

  - Affiliation/collaboration networks

- Biological networks

  - Metabolic networks

  - Protein-protein networks

  - Genetic regulatory networks

  - Drug interaction networks

  - Brain networks

    - Networks of neurons

    - Networks of brain functionality connectivity

  - Ecological networks

    - Food webs

    - Host-parasite networks

    - Mutualistic networks

Personally, my favorite networks are social networks (not social media companies), as they allow us to map out and understand a bit about the relationships between people, even on a massive scale. My *least* favorite networks are computer networks, as I am just not interested in subnet masking or different computer network architectures.

In this beautiful universe, graph theory, social network analysis, and network science give you the ability to investigate and interrogate relationships that many people don't even notice or recognize.

In this book, you will see the words graphs and networks used interchangeably. They are essentially the same structures but often for different purposes. We use graphs in *NetworkX* for network analysis. Those graphs are also called **sociograms** when they are visualized for social network analysis. Yes, that is confusing. You will get through this confusion, I promise. To make my life easier, I tell myself that they are the same things with different names, sometimes used for different purposes.

What is this network stuff?

Let's break things down a little further. I want to discuss the differences between graph theory, social network analysis, and network science separately. I am going to keep this very high level so that we can get to building as soon as possible. There are probably dozens of books available on Amazon if you want to dive deeper into any of these, and I personally probably have 10 or so books and will buy more as soon as I notice a new one is out.

Graph theory

There has been a lot of excitement about **graph theory** lately. I have noticed it the most in data science communities, but I have even seen database administrators and security professionals take an interest. Judging by the hype, one could assume that graph theory is something brand new, but it is actually *hundreds* of years old.

History and origins of graph theory

The history and origins of graph theory began in 1735, 286 years ago. Back then, there was a puzzle called *The Seven Bridges of Königsberg*, where the goal was to find a way to cross every one of the seven bridges without crossing any bridge twice. In 1735, Swiss mathematician Leonhard Euler proved that the puzzle was unsolvable. There was simply no way to cross each of the bridges without traversing a bridge twice. In 1736, Euler wrote a paper on this proof, which became the first-ever paper in the history of graph theory.

In 1857, an Irish mathematician, William Rowan Hamilton, invented a puzzle called the Icosian game that involved finding a special type of path, later known as a Hamiltonian circuit.

Almost 150 years after Euler's paper, in 1878, the term *graph* was introduced by James Joseph Sylvester in a paper published in *Nature*.

Finally, in 1936, the very first textbook on graph theory, *Theory of Finite and Infinite Graphs*, was written by Hungarian mathematician Dénes Kőnig, 201 years after Euler solved the seven bridges puzzle. In 1969, the definitive textbook on graph theory was written by American mathematician Frank Harary.

So, in other words, graph theory is not new.

Practical application of graph theory

Currently, there is much interest in graph theory as it relates to finding optimal paths. This is valuable knowledge due to its application in routing data, products, and people from point A to point B. We can see this in action when mapping software finds the shortest path between a source and a destination location. I also use paths in troubleshooting production database and software problems, so there are more applications than transportation.

We will work with the shortest paths throughout this book. Keep reading!

Social network analysis

Social network analysis has not received anywhere near the amount of hype that graph theory has, and I think that is a shame. Social network analysis is the analysis of social networks – the social fabric – that we are all a part of. If you study social networks, you learn to understand people's behaviors better. You can investigate who people interact with, who they hate, who they love, how drug addiction spreads, how white supremacists organize, how terrorists operate, how fraud is carried out, how malware spreads, how pandemics can be stopped, and on and on and on. There is math involved, but in my own research, I am most interested in uncovering social interaction. You can do social network analysis without writing out a single equation, so long as you understand what the techniques do, what assumptions they make, and how to tell whether they are working or not. Mathematically, I would say this is a friendly field, even if you are not mathematically inclined. However, mathematical prowess can be a superpower here, as it will allow you to dream up your own techniques that can be used to dissect networks and find insights.

History and origins of social network analysis

There is significant overlap between graph theory and social network analysis, as ideas such as the shortest path can be useful in a social network context for determining the number of handshakes it would take to get from where you are currently sitting to meeting the president, for instance. Or who would you have to meet in order to have a chance to meet your favorite celebrity?

However, in the 1890s, French sociologist David Émile Durkheim and German sociologist Ferdinand Tönnies foreshadowed the idea of social networks in their theories and research on social groups. Tönnies argued that social groups exist as ties that link individuals who share values and beliefs (community) or impersonal and instrumental social links (society). On the other hand, Durkheim gave a non-individualistic explanation, arguing that social phenomena arise when interacting individuals are part of a reality that is bigger than the properties of individual actors alone. I personally am intrigued by this idea, and I get a sense of this when I investigate where people are placed on a sociogram. Did the people create their position in a social network, or was some of it by chance, based on who they know and where their connections were already situated? For instance, if you are raised in certain parts of any country, you will likely inherit many of the sentiments shared by members of the community, but those sentiments may be much older than the living members of the community.

In the 1930s, major developments in social network analysis took place by several groups in psychology, anthropology, and mathematics working independently. Even 90 years ago, social network analysis was already a multi-disciplinary topic. In the 1930s, Jacob Moreno was credited with creating the first sociograms to study interpersonal relationships, and people were so intrigued by these that one was printed in the New York Times in 1933. He stated that *"Before the advent of sociometry, no one knew what the interpersonal structure of a group 'precisely' looked like."* Personally, whenever I visualize any social network that I have constructed for the first time, I feel that same intrigue. The network structure is a mystery until you first see it visualized, and it is always exciting to see it rendered for the first time.

In the 1970s, scholars worked to combine the different tracks and traditions of independent network research. Almost 50 years ago, people realized that independent research was going off in multiple directions, and there was an effort to bring it all together.

In the 1980s, theories and methods of social network analysis became pervasive among the social and behavioral sciences.

Practical application of social network analysis

Social network analysis can be used to study the relationship between any social entities. The goal of social network analysis is to understand how social entities and communities interact. This can be at the individual level or even at the international level.

This can even be used against literature. For instance, here is a social network I constructed from the book *Alice's Adventures in Wonderland*:

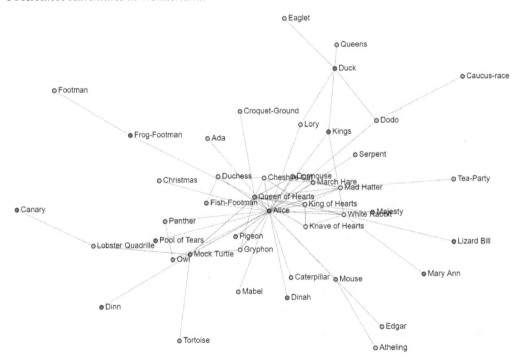

Figure 2.1 – Social network of Alice's Adventures in Wonderland

Here is a social network I constructed from the book *Animal Farm*:

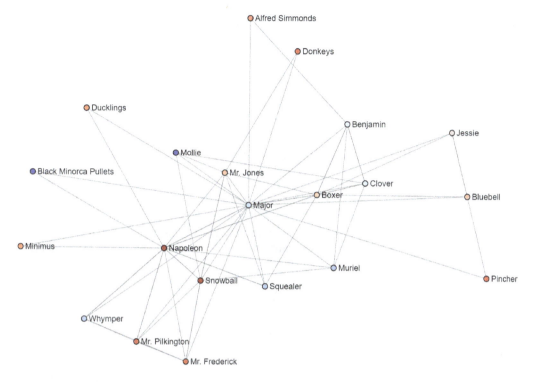

Figure 2.2 – Social network of Animal Farm

These are the original sociograms from when I first created them, so they could use a bit of work, and we'll work on cleaning them up a bit and analyzing them as part of this book.

I adore social network analysis more than graph theory or network science. I want to understand people. I want to stop bad people and make the world safer. I want to stop malware and make the internet safer. This powers everything for me, and so I spend a lot of my free time researching and doing social network analysis. This is my focus.

Network science

Network science is another fascination of mine. Network science is about understanding how networks form and the implications of how they form.

History and origins of network science

Like social network analysis, there is an overlap between the origins of graph theory and network science. To me, network science seems more about unifying various network disciplines and bringing everything under one roof, as data science is an attempt to unify various data disciplines. In fact, lately, I see network science emerging in use among data scientists, so I predict that, eventually, network science will be considered a subfield of data science.

In books I have read on network science, the same story of Euler's Seven Bridges is given as the origin of network science, which shows that graph theory is the origin of network science. However, there were a few key efforts in the 1990s that focused on mathematically describing different network topologies.

Duncan Watts and Steven Strogatz described the "small-world" network, where most nodes are not neighbors of each other but can reach each other in very few hops. Albert-László Barabási and Reka Albert discovered that small-world networks are not the norm in the real world and developed the scale-free network, where networks consist of a few hub nodes with many edges and many nodes with a few edges. This network topology would grow to maintain a constant ratio between the number of connections and all other nodes.

Practical application of network science

The practical application of network science is also the practical application of graph theory and social network analysis. I feel that graph theory and social network analysis are part of network science, even if techniques are specialized for each subject. I personally do not differentiate, and I don't think much about graph theory. I use techniques of graph theory and social network analysis when I do network science. When I do this via code, much of the mathematics of graph theory has been abstracted away, but I still very commonly use things such as shortest paths and centralities, the former from graph theory and the latter from social network analysis.

Resources for learning about network analysis

So, what is needed to begin a journey into network thinking? I'll give a few suggestions just to help you get started, but be aware that by the time this book has been published, some of this might already be a bit outdated and new technologies and techniques may have emerged. This is not a complete list. It is the bare minimum. The first thing that you need is a curious mind. If you are willing to investigate the hidden networks that make up our existence, then you already have the first prerequisite. Still, let me give you some more suggestions.

Notebook interfaces

I do all of my network analysis in Jupyter Notebooks. You can download and install Jupyter via Anaconda at: `https://docs.anaconda.com/anaconda/install`.

If you don't want to install Jupyter, you can also use Google Colab without any installation. You can find and immediately start using Google Colab at `https://research.google.com/colaboratory`.

Rather than a typical **Integrated Development Environment (IDE)** used for software development, a notebook interface allows you to write code in "cells" and then run them sequentially or re-run them. This is useful for research, as you can quickly experiment with data, doing exploratory data analysis and visualization in real time in your web browser. Here is what that looks like:

```
df = pd.read_csv(data)
df['book'] = df['book'].str.lower()
df['entities'] = create_entity_list(df)
df.head()
```

The Bible is contained in the data file, so this will load the data and output a preview of the first five verses of the Bible.

	book	chapter	verse	text	entities
0	genesis	1	1	In the beginning God created the heaven and th...	[God]
1	genesis	1	2	And the earth was without form, and void; and ...	[Spirit, God]
2	genesis	1	3	And God said, Let there be light: and there wa...	[God]
3	genesis	1	4	And God saw the light, that it was good: and G...	[God, God]
4	genesis	1	5	And God called the light Day, and the darkness...	[God, Day, Night]

Figure 2.3 – pandas DataFrame of the Bible

Now, let's see in which verses `Eve` is mentioned in `genesis`:

```
entity_lookup('Eve', book='genesis')
```

This will display a preview of the two verses in which `Eve` was mentioned:

	book	chapter	verse	text	entities
75	genesis	3	20	And Adam called his wife's name Eve; because s...	Eve
80	genesis	4	1	And Adam knew Eve his wife; and she conceived,...	Eve

Figure 2.4 – Eve mentions in Genesis

Let's see how many times `Eve` is mentioned by name in the entire Bible:

```
entity_lookup('Eve')
```

Again, this will display a preview of verses:

	book	chapter	verse	text	entities
75	genesis	3	20	And Adam called his wife's name Eve; because s...	Eve
80	genesis	4	1	And Adam knew Eve his wife; and she conceived,...	Eve
28992	2 corinthians	11	3	But I fear, lest by any means, as the serpent ...	Eve
29729	1 timothy	2	13	For Adam was first formed, then Eve.	Eve

Figure 2.5 – Eve mentions in the entire Bible

I have loaded the Bible dataset that I created and done a bit of wrangling and enrichment, and then I have created two additional cells that look for `Eve` in the book of Genesis and then in the entire Bible. If you look closely, you can see that I accidentally set `df` as a global variable, which is not good. I should fix that code. Oh well, next time. It works.

A notebook allows you to experiment with data and get really creative. There are pros and cons to notebooks, one of which I previously mentioned.

The pros are that you can very quickly and easily experiment on datasets, building out your code for loading, preprocessing, enrichment, and even **Machine Learning** (**ML**) model development. You can do a LOT with a notebook, very quickly. It is my favorite tool for initial exploration and development.

The cons are that notebook code has a deserved reputation for being quick and dirty code, not something that really belongs in production. Another problem is that notebooks can get disorganized, with notebooks on our laptops and Jupyter servers (SageMaker, Kubeflow, and so on), so things can quickly get out of hand and confusing.

I recommend using them to explore data and get innovative, but once you have something that works well, take that code into an IDE and build it out properly. My preferred IDE is currently Visual Studio Code but use whatever you are comfortable with. IDEs will be defined in the next section.

You can download Visual Studio Code at `https://code.visualstudio.com/download`.

IDEs

An IDE is a piece of software that is used for writing code. Examples of IDEs are Visual Studio Code, Visual Studio, PyCharm, IntelliJ, and many more. There are many different languages, and they often support multiple programming languages. Try a few and use whatever is most comfortable. If it slows you down too much for too long, it's not the right one. However, there is typically a bit of a learning curve with any modern IDE, so be patient with yourself and with the software. If you want to learn more about available IDEs, just google `popular IDEs`. When getting started, I recommend setting some time aside to really learn how to use your chosen IDE. Watch videos and read the documentation. I typically do this every few years, but absolutely at the beginning.

Network datasets

You can't do network analysis without data. You either need to create data, or you need to go find some. Creating network datasets can be easier, simpler, and less manual than creating other types of datasets (for instance, ML training data), and it can be a good way to build domain knowledge quickly. Our own interests and curiosities can be more enjoyable to investigate than using someone else's dataset. But so that you know, there are network datasets out there that you can start with while you get comfortable with the network analysis toolset:

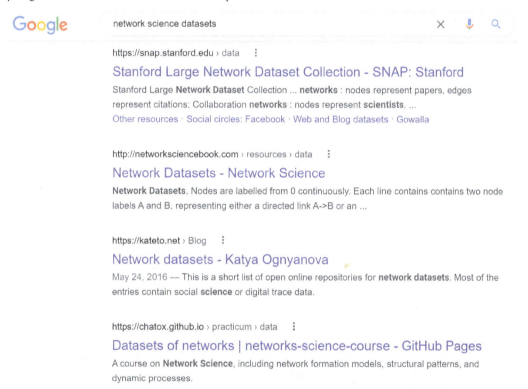

Figure 2.6 – Google search: network science datasets

All of them look a bit neglected and basic, other than the top link. They appear not to have been updated in several years.

Searching for social network analysis datasets gives slightly more interesting-looking datasets, which makes sense, as social network analysis is older than network science. However, the top link is the same. If you want some pre-made data, take a look at the Stanford website as well as `data.world`.

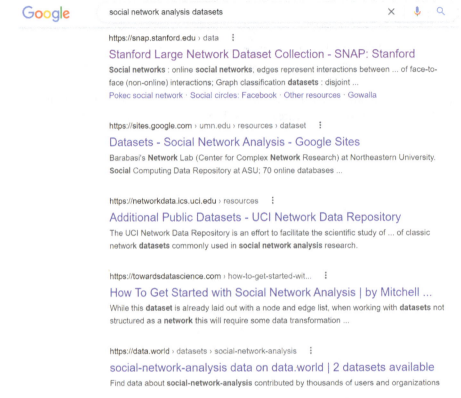

Figure 2.7 – Google search: social network analysis datasets

Kaggle datasets

Kaggle is another decent place to look for data for just about anything. Kaggle is occasionally given unfair treatment by some people, who snub their noses at those who use Kaggle datasets because the data isn't dirty like in the real world, but there is little difference between Kaggle data and the preceding datasets. They have all been scrubbed clean and are ready for you to practice with ML or network analysis tools. There are so many interesting datasets and problems to solve on Kaggle, and it is a great place to learn and build confidence. Use what helps you.

NetworkX and scikit-network graph generators

In this book, I will explain how to use the Python libraries *NetworkX* and *scikit-network*. I use NetworkX for graph construction and scikit-network for visualization. I do that because scikit-network has no real network analysis capabilities, but it renders network visualizations in **Scalable Vector Graphics** (**SVG**), which are much faster and bettee-looking than NetworkX visualizations. Both of these libraries also contain generators for several good practice networks. If you use them, you won't get familiar with how to create a network from scratch, but it's an OK place to start if you just want to experiment with a pre-built network:

- NetworkX: `https://networkx.org/documentation/stable/reference/generators.html`

- scikit-network: `https://scikit-network.readthedocs.io/en/latest/tutorials/data/toy_graphs.html`

My recommendation is that you should just ignore the scikit-network loaders and get familiar with creating networks using NetworkX. It is just far superior, in my opinion. At this time, scikit-network is in no way a replacement for NetworkX, except for visualization.

Creating your own datasets

In this book, I am going to show you how you can create your own datasets. Personally, I feel that this is the best approach, and this is what I do practically every time. Usually, I get an idea, write it on my whiteboard or in a notebook, figure out how to get the data, and then get to work. This typically takes a few days, and I'll usually use an approach like this:

1. Come up with an idea.
2. Run to a whiteboard or notebook and write it down.
3. Go sit outside with a notebook and daydream about how I can get the data.
4. Set up some web scrapers.
5. Download the data after a day or two of scraping.
6. Extract entities from sentences and build an edge list dataframe.
7. Construct the network.
8. Dive into network analysis.
9. Visualize the things that seem interesting.
10. Play with ego graphs to look into ego networks for additional perspective.

This is definitely a process, and it seems like a lot of steps, but over time, it all blends together. You have an idea, get data, do preprocessing, and then analyze and use the data. In a way, that isn't very different from other data science workflows.

In this book, I will explain how to take raw data and convert it into network data.

There are lots of options for where you can get network data. None of them are wrong. Some of them just skip steps and make things easier so that you can practice with tools and techniques. That is perfectly fine. However, you should aim to become self-sufficient and comfortable working with dirty data. Data is really never clean, even if someone seems to have done a lot of cleanup activity. There's usually more to do, even in my own datasets from my own GitHub. Text is messy. You will need to get comfortable playing in the dirt. For me, that is my favorite part—taking gnarly text data and converting it into a beautiful and useful network. I love it.

So, get familiar with everything we've talked about and go explore. Maybe even consider making the first network data repository that somehow succeeds in bringing everything together if you are feeling really ambitious.

NetworkX and articles

Another thing I have found useful for learning about network science is the article links that are posted directly on the NetworkX documentation website. When they describe their functions, they typically provide a link to an article about the technique that was used to build the function. Here is an example:

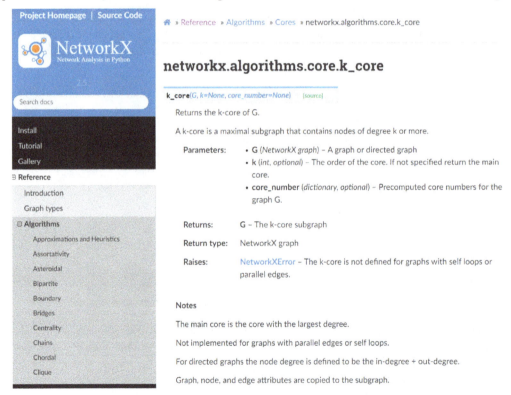

Figure 2.8 – NetworkX documentation: k_core algorithm

If you go to the NetworkX page for k_core, you can see clear documentation about the function, parameters, what it returns, and some notes. This isn't even the best part. The best part is buried a little deeper. Scroll down, and you will often see something like this:

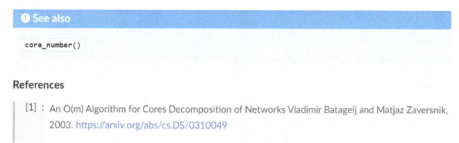

Figure 2.9 – NetworkX documentation: k-core article reference

Nice! There is an arXiv link!

It is so useful to be able to read about the research that led to k_core. It is also somewhat of a relief, to me personally, to find out that so much stuff is actually quite old, for instance, what I mentioned in terms of **Natural Language Processing (NLP)** history. We often hear that it is impossible to keep up with data science because things move so fast. That is only true if you are distracted by every shiny new thing. Focus on what you are trying to do, research ways to do what you want to do, and then build it. You shouldn't only use newer techniques. New does not always, or even usually, mean better.

In the case of k_core, I was looking for a way to quickly throw away all nodes with one edge. In my first networks, I would write way too many lines of code to do that, trip over a list comprehension, or just write some nasty code that worked. However, I could have just done nx.k_core(G, 2). That's it. Problem solved.

So, read the documentation. Look for academic journal links if you want to learn more, and then build and build and build until you understand the technique.

You will find the same kinds of links for centralities or other topics on NetworkX. Just go explore. It's good reading and inspiring stuff.

Very nice! This sets us up to learn more about the origins and use of centralities in network analysis.

Common network use cases

As I did in *Chapter 1, Introducing Natural Language Processing*, I will now also explain some of my own favorite use cases for working with network data. I mentioned at the beginning of the chapter that there are many different kinds of networks, but I personally prefer working with social networks and what I call *dataflow networks*.

Here are some of the uses I have for working with network data:

- Mapping production dataflows

- Mapping community interactions

- Mapping literary social networks

- Mapping historical social networks

- Mapping language

- Mapping dark networks

I will start with dataflow networks, as that was the first use case I realized for network data and something that revolutionized how I work.

Mapping production dataflow

As mentioned, this was the first idea that I had for my own use of network data. I have worked in software for over 20 years, and I have spent a significant amount of time "dissecting" production code. This isn't for fun. This has serious use. In the past, I had been asked to "uplift" old production systems, which means taking an old production system, figuring out everything that runs on it (code, databases, and file reads/writes), and migrating everything over to a new server so that the old server could be retired. We call this the *de-risking* of production systems. In the end, this results in a lightning-fast new production system because this is usually old code placed on very new infrastructure.

In my old way, I would take an inventory of production systems by starting with `cron`, the orchestration of how things run on a server. Starting at `cron` allows you to get to the root of a process and see what scripts call additional scripts. Often, if you map this out, you'll end up with something like this:

- cron:

  - `script_a.php > script_a.log`

  - `script_b.php > script_b.log`

  - `script_c.php > script_c.log`

  - `script_c.php uses script_d.php`

  - `script_c.php uses script_e.php`

And on, and on, and on. That's not what `cron` actually looks like, but that shows the direction of script calls and data writes. So, in the previous example, the code does the following:

- `script_a.php` writes to the `script_a.log` logfile

- `script_b.php` writes to the `script_b.log` logfile

- `script_c.php` writes to the `script_c.log` logfile
- `script_c.php` launches another script, `script_d.php`
- `script_c.php` launches another script, `script_e.php`

I used to do this by hand, and it was very tedious and exact work, as it can be catastrophic if you miss anything because it won't be migrated or tested on the new server, and that is a painful thing to discover on migration day. On a server with a couple of hundred scripts, it would take a few weeks to even build a script map, and a full map of the entire input and output of the entire server would take much longer than that. From the preceding example, a script map would look like this:

- cron:

  - `script_a.php`

  - `script_b.php`

  - `script_c.php`

    - `script_d.php`

    - `script_e.php`

In a production environment, this can go very deep and be very nested. In 2017, I found another way to do this, using networks, and the script map would build itself using the nodes and edges constructed in the graph. This could be an exceptionally long discussion that I could write an entire book on – and I have thought about doing just that – but to save time, the new method takes roughly 10% of the amount of time as the old way and catches everything, as long as I am thorough in my source code analysis. I also do not have to maintain a mental image of what is going on in the server, as I have a graph visualization that I can visually inspect and a network that I can interrogate. Working 10 times faster, I can uplift many more servers in the same amount of time it used to take me to do a single one. I can also use this to quickly troubleshoot production problems once the network is built. Problems that used to take days to troubleshoot, now I can often do in minutes.

Another nice side effect is that if you want to re-engineer software and you use this approach, you can understand the inputs and outputs of everything on a server, and you can use that to find inefficiencies and come up with ideas on how to better build the software on a new platform. I have also done this, replacing SQL Server databases with Spark dataflows, and very successfully.

Mapping community interactions

These days, I no longer uplift production systems and have moved on to more exciting work, so everything else I am going to describe comes from my own independent research, not from things I do at work. For my own research, I do a lot of social media scraping and analysis, as it gives me a way to understand the world around me. One thing I like to understand is how people in certain groups interact with each other and with others outside of their group.

For instance, I scrape dozens of data science-related Twitter accounts so that I can keep an eye on the latest occurrences in the data science and scientific communities. After initially scraping a few accounts, I extract user mentions and map all the interactions between them. This helps me find additional users that are worth scraping and add them to my scrapers, and then I repeat extracting user mentions and map all the interactions between them as well. I do this over and over and over, constantly building out social networks for anything I am interested in—data science, natural science, art, photography, and even K-pop. I find analyzing communities intriguing, and it's one of my favorite things to do. This is one topic I will spend a great deal of time covering in this book because I have seen how useful this can be to research, and I have even used the results of this scraping to build highly customized ML training data so that I can automatically find even more of what I want! The results are a beautiful combination of NLP, social network analysis, and ML!

This is also useful for keeping an eye on trends and topics of discussion within communities. For instance, if there is a big data science conference coming up, there will suddenly be a lot of chatter on the topic, probably sharing similar hashtags. These hashtags can also be found by creating user-hashtag networks rather than user-user networks. This is excellent for watching trends come and go, especially if you add a time element to your network data.

Mapping literary social networks

Another one of my favorite uses of networks is to take a literary piece, such as *Alice's Adventures in Wonderland* or *Animal Farm,* and map out the social network of the book. I will discuss this in detail as part of this book. It is a bit more involved than what I do for social media, as the previous technique can be done with regex alone, and this technique requires NLP part-of-speech tagging or named-entity recognition. However, as a result, you end up with a social network of any piece of text that you can visually inspect to gain a better understanding of how characters interact.

Mapping historical social networks

You can also use NLP (part-of-speech tagging and named-entity recognition) approaches to map out historical social networks, as long as you have text about an event or some kind of data that shows relationships between people. For instance, it is simple to build the social network of the entire Bible by doing what I described with literary social networks. There is quite a lot of cleanup involved, but the results are breathtaking if you are thorough. I would love to see researchers do this with other historical texts, as it may give us additional insights that we have never discovered before.

Mapping language

Another cool use for networks is mapping language itself. This is something I discovered while playing around with the book *Pride and Prejudice* by Jane Austin. I wanted to see what would happen if I built a network of the interactions between every word in every sentence of her book. For instance, let's take the following sentence from *Pride and Prejudice*:

> *It is a truth universally acknowledged, that a single man in possession of a good fortune, must be in want of a wife.*

Let's map this out using word sequence for setting up word relationships. If a word follows another word, then we say that there is a relationship between the two. Think of a relationship as an invisible line connecting two things. For the previous quote, we could map it out as such:

- it -> is
- is -> a
- a -> truth
- truth -> universally
- universally -> acknowledged
- acknowledged -> that
- that -> a
- a -> single
- single -> man
- man -> in
- in -> possession
- possession -> of
- of -> a
- a -> good
- good -> fortune
- fortune -> must
- must -> be
- be -> in
- in -> want
- want -> of

- of -> a

- a -> wife

In networks, we call the connections between nodes **edges**. **Nodes** are the things that we are connecting. In this case, this is a word network, so words are nodes, and the connections between words are called edges. You can think of edges as lines between two things.

In the quoted sentence, you may notice that a has an edge with `truth`, `single`, `good`, and `wife`. `in` also has two edges, with `possession` and `want`.

Here is another example using a song by the Smashing Pumpkins:

Despite all my rage, I am still just a rat in a cage.

You could also map it out as follows:

- despite -> all

- all -> my

- my -> rage

- rage -> i

- i -> am

- am -> still

- still -> just

- just -> a

- a -> rat

- rat -> in

- in -> a

- a -> cage

The letter a has an edge with both `rat` and `cage`.

If you do this with an entire book, you end up with a super-dense network, and it is nearly worthless to visualize it; however, analyzing the network is not worthless. If you look at the outer nodes of *Pride and Prejudice*, you will find Jane Austin's least used words, which often make an amazing vocabulary list. If you look at the core words, they are typically bridge words, such as "a," "the," and "of."

I also used this same technique for some research to see whether I could visually tell the difference between AI-generated text and human-generated text. Guess what? You can. If you use ML to generate text, it will choose the next word not because it is the most fitting but because it has the highest probability of being used. That's great when you are trying to predict the next word in the sentence

"Jack and Jill went up the _____", but that is not great when you actually want to use it to create something truly original. We, humans, are more nuanced. If I wanted to transform that sentence into something really morbid, I could complete the sentence with the word "chimney." I really doubt ML would choose chimney.

And visually, the networks look different between AI-generated text and human-generated text. When I did this back in 2019, the AI-generated network looked really jagged on the outside, with a much denser core, because it would choose higher-probability words. With humans, we are more nuanced, and the outer edge appeared softer. I was able to tell the difference between the two visually. Would more recent AI appear different? I encourage you to try this out if it interests you.

Mapping dark networks

This last one is a bit dark. Sometimes, when bad things happen in the world, I want to know what groups of people are planning. Dark networks are social networks of people who wish to do others harm, but there are different kinds of dark networks (related to crime, terror, hate, drugs, and so on). Even dangerous people will rally around certain social media hashtags, so they occasionally hint at what is about to happen. I will not go into great detail on the level of intelligence that can be gathered, but you can do what I discussed under *Mapping community interactions* and use that to keep an eye on things or find other people of interest. You can also use that to report social media posts that violate community guidelines and even escalate to law enforcement if you see something specific. Just know that if you do this kind of **Open Source Intelligence (OSINT)** gathering, it can have a nasty effect on your mood and mental health, so tread lightly. However, it can also be excellent research.

> **Important note**
>
> When I am using this for OSINT, I typically start with a few people or things of interest and then gradually map out the social network that exists around them. However, this is not limited to dark networks. You could use this same technique for offense, defense, or just to follow your own curiosity. It is one useful way to study the interactions between people and organizations.
>
> There are many uses for OSINT, and entire books have been written on the subject. However, the goal of OSINT is to use publicly available data to better understand or monitor something. For instance, you could use OSINT to research a company that you were interviewing at. Who is the owner? Who is interviewing me? Has the company been in the news? Was it a positive or negative story? What is the general consensus of the public about this company? Is it a legitimate company? What are ex-employees saying about it? What do they say on social media? Are most comments positive, negative, or neutral? There is so much you can learn about anything with the right set of questions and access to the internet.

Market research

Let's consider a hypothetical situation: you have created a new product and would like to identify people who may find your product useful. I have done exactly this for my own needs.

I have used social network analysis to build a map of people who are part of a community and have used the results for market research. For one topic of interest, I can see thousands of researchers from around the world, and if I were to create a product for that community, it would take little work to develop an outreach campaign that would effectively drive interest in my product. For this, I only used Twitter data and was able to identify thousands of people I may want to approach after only a few days of scraping and analysis.

Finding specific content

This part is kind of a chicken-and-egg scenario. You need social media text in order to extract the social network data that exists in the text, and then you can use the social network to identify other accounts to scrape, further improving your data feeds. This is an iterative process. Identify, scrape, analyze, identify, scrape, analyze, and so on. You can further improve the analysis part by building NLP classifiers off the text you have scraped. Social network analysis will help you find important accounts, but your NLP classifiers will help you filter the noise to get relevant and focused content.

Creating ML training data

As mentioned, you can use the data feeds you scrape to create NLP classifiers, and you can use those classifiers to help you filter out noise, resulting in more focused content. You can then use that focused content to build even better classifiers. However, you must first create your initial classifiers, and the feeds that you create using a combination of NLP and network analysis will give you the rich, filtered data that you need to create useful training data. There is a process to this. It is time-consuming and takes a lot of work. I will explain this in *Chapter 14, Networks and Supervised Machine Learning*.

Specific and focused content feeds into classification training data, and trained classifiers will help identify additional useful content.

Advanced network use cases

In *Chapter 1*, *Introducing Natural Language Processing*, I specified several advanced use cases for NLP, such as language translation and text generation. However, while thinking about network analysis, my mind immediately asked, what would an advanced network use case even mean? This is all pretty advanced stuff. With NLP, you have simple tasks, such as tokenization, lemmatization, and simple sentiment analysis (positive or negative, hate speech or not hate speech), and you have advanced tasks. With networks, I can think of three potentially advanced use cases:

- Graph ML

- Knowledge graphs

- Recommendation systems

However, I don't think of any of them as all that advanced. I think of them as just having different implementations from other things I have mentioned. Furthermore, just because something is more technically challenging does not make it advanced or more important. In fact, if it is more difficult and returns less useful results, that's not ideal. That's a time sink.

Graph ML

I have worked on graph ML projects and have primarily found that there are two approaches: use graph measures (centralities, local clustering coefficients, and so on) as features in training data or feed graph representation data directly to ML and let it figure things out. Graph metadata can be a powerful enrichment for your training data as it can give an additional perspective for a model to hook into, so I think this is really promising.

However, in the past, when I have seen people attempting to use graph data directly in ML, often the graph knowledge was absent or weak. I wouldn't recommend this approach until you are comfortable doing network analysis and ML separately. In ML, input data is easily as important (or more) than the model itself, so knowledge of networks should be present, but that is not always the case. Domain knowledge is important in data science. However, it is still extremely interesting stuff, so definitely check it out.

Recommendation systems

Recommendation systems are interesting, though I haven't spent much time researching them; one concept is that if two people like similar things, they may also like other things that neither yet have in common. For instance, if you and I both like the bands Soundgarden, Smashing Pumpkins, and Nirvana, and I like The Breeders, and you like Stone Temple Pilots, it is quite likely that I'll also like Stone Temple Pilots and you'll also like The Breeders. If you want to explore recommendation systems, I encourage you to dive into the research. It's just not one of my interests, as I primarily use networks for social network analysis, market research, and OSINT.

However, there are downsides to recommendation systems that I would like to point out. These systems recommend things that we would probably like, but the results aren't usually all that unexpected. Musically, it makes sense that I would probably like Stone Temple Pilots and that you'd like The Breeders, but I am personally more excited when I fall in love with something completely unexpected. I am hopeful that in the future, our recommendation systems will recommend food, music, and TV shows to us that we would not discover otherwise. I don't want to live my life only experiencing similar things. I want the unexpected, as well.

Also, when we are only given things that we expect, we often end up reinforcing not-so-great things. For instance, if social media companies only show us stories that they think that we would like to see or that we are likely to engage with, we end up in an echo chamber, and this causes us to primarily associate with people who are just like us, who talk about the same things, and who agree upon the same things. That's neither educational nor constructive. That causes polarization and can even lead to violence. It becomes a dangerous problem when people will only associate with those who are just like them and when they begin to demonize others who disagree with them, yet this happens all the time, and we are susceptible.

I'll leave it up to you to decide what is a common versus advanced use of networks. In my mind, they're all advanced, the implementation might just be different.

Getting started with networks

If you want to jump into your first original network analysis project, you need to first think up something that you are curious about. With social network analysis, you are often looking to build a sociogram, a visual map of how humans interact. So, for your first network project, you could ask something such as the following:

- How do the characters from <book> interact with each other?

- Do the different animals in the book *Animal Farm* only interact with the same type of animals? Do humans only interact with certain types of animals, or do they interact with all types of animals?

- What does the Twitterverse of my own town look like?

- What does a network visualization of ingredients to meals look like? How does this differ for different regions in the world? What does a network visualization of ingredients to region look like?

- What does the Twitter social network that exists around a certain politician look like? What does the friend map look like? What does the enemy map look like?

- If I build an example dark network, what would be the optimal way to shatter it into pieces irreparably so that it would never be able to re-form if under constant attack?

- If I build a network of my own infrastructure, and if I were to identify the structural weaknesses that could cause it to shatter into pieces if exploited, how could I best protect it now to prevent those types of attacks from disrupting the network?

And these are just a few that I thought up in the last few minutes. The more you experiment and play with networks, the easier it will become for you to come up with your own research ideas, so just start with something simple and then become more ambitious over time. Some of these probably sound pretty complicated, but as long as you can get the data that you need, the network construction, analysis, and visualization are really not too difficult and have a lot of overlap.

In short, you need to find something that you are interested in and identify a relationship between those things. This could be as simple as who knows who, who likes who, who trusts who, who hates who, who works with who, what ingredients go into what meals, what infrastructure communicates with other infrastructure, what websites link to other websites, or what database tables can join against other database tables.

For my own independent research, I typically use Twitter data, as in my opinion, it is an NLP and network goldmine. And much of my work follows a repeatable process:

1. Come up with a research topic.
2. Find a few Twitter accounts that are related to that topic.
3. Scrape the accounts.
4. Analyze the scraped data:

    - Extract network data.
    - Extract additional users to scrape.

5. Repeat *steps 3 and 4* as far as you want to go.

As mentioned, it's repeatable. I'll explain how to get started with scraping in *Chapter 6*, *Graph Construction and Cleaning*, and *Chapter 7*, *Whole Network Analysis*. Once you learn how to get data via scraping and **Application Programming Interfaces** (**APIs**), you will have access to more data than you can ever use. And for social network analysis, you will be able to follow the preceding example. Go slowly, and it will eventually become natural.

Example – K-pop implementation

My daughters are into K-pop music. Frequently, during dinner, I'd hear about groups such as BLACKPINK, TWICE, and BTS, and I had no idea what my girls were talking about. Then, one day, I was trying to think up a topic of research for my LinkedIn challenge *#100daysofnetworks*, and I thought I should do something with K-pop, as I could relate with my daughters if I knew a bit more about it, and it'd give me a way to show my daughters a little more about data science. So, I came up with a process to turn this idea into a reality:

1. I decided that I wanted to do some research into what the social network looked like around many of the most famous K-pop artists, and this should include music hubs and fan clubs.

2. I did about 30 minutes of Twitter lookups, hoping to identify a couple of dozen K-pop groups, music hubs, and fan clubs. I found about 30 of them and wrote them in a notebook.

3. I added these 30 accounts to my Twitter scraper and then combined the data into a feed, `kpop.csv`. This makes it much easier for me to do all of my network data extraction and then analysis, with everything in one file. My scrapers run 24/7/365, so I always have fresh data available to explore.

4. After getting a few days' worths of scraped data, I analyzed the feed, extracted network data (user mentions and hashtags), and identified a couple dozen more accounts to scrape.

5. I repeated *steps 3 and 4* for about a month, and I now have a nice Twitter-centric social network of K-pop artists, music hubs, and fan clubs.

And now that I am currently actively scraping 97 K-pop-related accounts, I have fresh new media all the time for my daughters to enjoy, and I have interesting network data to use for further improving my skill.

So, for your own network research, find something that you are interested in and then get to work. Choose something that you are interested in. Research should not be boring. This book will show you how to take your research curiosity and turn it into results. We will go into graph construction, analysis, and visualization after we get through the next couple of chapters.

Summary

We covered a lot of material in this short chapter. We discussed the confusion around the word *network*, went into the history and origins of graph theory, social network analysis, and network science, discussed resources for learning and practice, discussed some of my favorite network use cases, and finished by explaining how you can start formulating your own network research.

I hope this chapter gave you a rough idea of what all of this network stuff is. I know I did not go into great detail on the origins, and I mostly talked about social network analysis, but that is because that is my area of interest. I hope you now understand what networks can be used for, and I hope you understand that I have only scratched the surface. My goal was to ignite your curiosity.

In the next chapter, I will explain the tools used for NLP. We are going to gradually move past theory and into data science.

Further reading

- Barabási, A.L. (2014). *Linked*. Basic Books.
- Kurzweil, R. (2012). *How to Create a Mind*. Penguin Books.
- Newman, M. (2018). *Networks*. Oxford University Press.

3
Useful Python Libraries

In this chapter, I'm going to introduce several of the Python libraries we will be using in this book. I will describe what they are and how to install them, and give examples of a few useful things you can do with them. You do not have to memorize all of the capabilities of every Python library that you use, or thoroughly understand the internals of every single function that you use. It is important that you understand what libraries are available, what overall capabilities they have, and what critically important timesavers are in the library (this will keep you coming back for more). Everyone's use case is different, and there is no way to memorize everything. I recommend that you internalize this point as quickly as possible, and learn what you need to when you need it. Learn about the internals in as much depth as you need.

To keep this organized, I am separating software libraries by category. Here are the libraries we will discuss:

Python Library	Category
pandas	Data Analysis and Processing
NumPy	Data Analysis and Processing
Matplotlib	Data Visualization
Seaborn	Data Visualization
Plotly	Data Visualization
NLTK	Natural Language Processing
spaCy	Natural Language Processing
NetworkX	Network Analysis and Visualization
Scikit-Network	Network Visualization (Better)
scikit-learn	Machine Learning
Karate Club	Machine Learning (Graph)
spaCy (Repeat)	Machine Learning (NLP)

Figure 3.1 – A table of Python libraries for NLP

It's useful to break the libraries down this way, as it is logical. We need to first collect, process, and analyze data before we should do anything else. We should visualize the data as we analyze it. Certain libraries specialize in NLP. Other libraries specialize in network analysis and visualization. Finally, there are different libraries that are useful for different kinds of ML, and even non-ML-focused libraries often have ML capabilities, such as spaCy.

I am going to keep this very high level, as we will actually be using these libraries in upcoming chapters. This chapter covers the "what" and "why" questions about given libraries. Other chapters will cover how they may be used.

One final point before moving on: you do not need to memorize every aspect of any of these Python libraries. In software development and data science, skills are acquired piece by piece. Learn what is useful to you now, then if a library proves to be useful, learn more. Don't feel guilty for only knowing small pieces. Knowledge is accumulated over time, not all at once.

Our sections for this chapter will be the following:

- Using notebooks
- Data analysis and processing
- Data visualization
- NLP
- Network analysis and visualization
- ML

Technical requirements

This chapter will deal with a lot of the resources we will be using throughout the book.

All of the code can be found in the GitHub repository: https://github.com/PacktPublishing/Network-Science-with-Python.

Using notebooks

It is often easiest – and very useful – to do data analysis and prototyping using what we often affectionately just call **notebooks**. **Jupyter** defines the Jupyter notebook as a web-based interactive computing platform. I like that simple definition. Notebooks are essentially a series of "cells" that can contain code or text, which can be run individually or sequentially. This allows you to write code in a web browser, run the code while in the web browser, and see immediate results. For data analysis or experimentation, this immediate feedback is useful.

In this book, we use Jupyter Notebook. I recommend downloading and installing it from the Anaconda website. You can do so at https://www.anaconda.com.

In Jupyter, you can run code and see the immediate results of that code, whether the output be text, numeric, or a data visualization. You will see a lot of notebook use in this book, so I will keep this short.

Google Colab is another option for working with notebooks, and you don't have to install anything to use it. This can be an even easier way to work with notebooks, but it has its advantages and disadvantages. I recommend that you learn how to use both and choose the one that you like best, or that allows you to work well with others.

You can check out Google Colab at `https://colab.research.google.com`.

Next, let's explore the libraries we'll use for data analysis.

Data analysis and processing

There are a number of useful libraries for working with data, and you will want to use different libraries and techniques at different points of the data life cycle. For instance, in working with data, it is often useful to start with **Exploratory Data Analysis (EDA)**. Later on, you will want to do cleanup, wrangling, various transformations for preprocessing, and so on. Here are some of the available Python libraries and their uses.

pandas

pandas is easily one of the most important libraries to use when doing anything with data in Python. Put simply, if you work with data in Python, you should know about pandas, and you should probably be using it. You can use it for several different things when working with data, such as the following:

- Reading data from an assortment of file types or from the internet

- EDA

- **Extract, Transform, Load (ETL)**

- Simple and quick data visualizations

- And much, much, more

If there is one Python library that I would recommend to everyone on this planet – not just data professionals – it is pandas. Being able to make short work of data analysis outside of spreadsheets is liberating and powerful.

Setup

If you're working in Jupyter or Google Colab notebooks, it's very likely that pandas is already installed and that you can just start using it. However, if you need to follow the steps, you can follow the official installation guide: `https://pandas.pydata.org/docs/getting_started/install.html`.

Once it is installed, or if you suspect it is already installed, simply run this statement in your favorite notebook:

```
import pandas as pd
```

If Python doesn't complain about the library not being installed, you are all set.

If you want to see what version of pandas you are using, run this statement:

```
pd.__version__
```

I can see that I am running version 1.3.4.

Starter functionality

Reading data is my preferred way to help people get started with pandas. After all, data work is pretty boring if you don't have any data to play with. In pandas, you typically work with a DataFrame, which resembles a data table or spreadsheet. A pandas DataFrame is a data object consisting of rows and columns.

To read a CSV file into a pandas DataFrame, you would do something like this:

```
data = 'data/fruit.csv'
df = pd.read_csv(data)
```

The data that I want to use is in a subdirectory of the current directory called data, in a file called fruit. csv.

Once it's read into a DataFrame, you can preview the data:

```
df.head()
```

This will show a preview of the first five rows of the DataFrame.

	fruit	count
0	apple	3
1	orange	4
2	kiwi	1
3	banana	9
4	strawberry	6

Figure 3.2 – Simple pandas DataFrame

You can also read other datasets into `pandas`. For instance, if you already have `scikit-learn` installed, you could do this:

```
from sklearn import datasets
iris = datasets.load_iris()
df = pd.DataFrame(iris['data'], columns=iris['feature_names'])
df.head()
```

That gives us *Figure 3.3*:

	sepal length (cm)	sepal width (cm)	petal length (cm)	petal width (cm)
0	5.1	3.5	1.4	0.2
1	4.9	3.0	1.4	0.2
2	4.7	3.2	1.3	0.2
3	4.6	3.1	1.5	0.2
4	5.0	3.6	1.4	0.2

Figure 3.3 – pandas DataFrame of the iris dataset

`pandas` doesn't only read from CSV files. It is quite versatile, and it is also less error-prone than other ways of loading CSV files. For instance, I often find that **Spark** glitches out when reading CSV files, while `pandas` will have no problem.

> **What is Spark?**
>
> Spark is a technology that is often used for processing large amounts of data. Spark is very useful for "big data" workloads. At a certain point, large amounts of data can become too much for tools such as `pandas`, and it is useful to know more powerful tools, such as Spark.

Here are some of the other `pandas` `read_*` functions:

- `pd.read_clipboard()`
- `pd.read_excel()`
- `pd.read_feather()`
- `pd.read_fwf()`

- `pd.read_gbq()`
- `pd.read_hdf()`
- `pd.read_html()`
- `pd.read_json()`
- `pd.read_orc()`
- `pd.read_parquet()`
- `pd.read_pickle()`
- `pd.read_sas()`
- `pd.read_spss()`
- `pd.read_sql()`
- `pd.read_sql_query()`
- `pd.read_sql_table()`
- `pd.read_stata()`
- `pd.read_table()`
- `pd.read_xml()`

This looks like a lot, and I personally do not have all of these memorized. I typically only use a handful of these, such as `read_csv`, `read_json`, and `read_parquet`, but `read_html` is occasionally useful at a pinch as well.

For me, EDA and simple visualizations are what got me hooked on `pandas`. For instance, if you wanted to draw a histogram, you could do this:

```
df['petal width (cm)'].plot.hist()
```

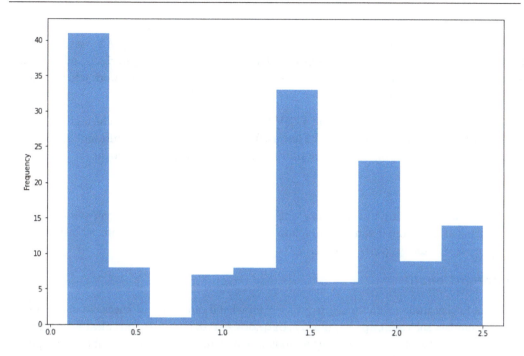

Figure 3.4 – Petal width histogram

pandas can do much more than what I've shown. There are countless blog posts and books on doing data analysis with pandas. Whole books have been written about using pandas. I encourage you to buy a few and learn as much about this library as you can. The more you learn about pandas, the better you will be at working with data.

Documentation

You can learn more about pandas by visiting this website: https://pandas.pydata.org/docs/getting_started/overview.html.

You can use this book as a reference to get more comfortable working with pandas, but there are also several online guides that will give hands-on practice, such as this: https://pandas.pydata.org/docs/getting_started/intro_tutorials/01_table_oriented.html.

NumPy

NumPy (Numeric Python) advertises itself as the fundamental package for scientific computing in Python, and I can't argue with that. Where pandas is excellent for doing data analysis against DataFrames, NumPy is more general-purpose and consists of a wide variety of mathematical functions and transformations.

But what's the difference between it and `pandas`? That's actually a good question because both libraries are often used together. Where you see `pandas`, you will often also see `NumPy`. After several years of using both, the way things typically happen is I'll try to do something in `pandas`, find out that it's not possible, and then find out that it's possible in `NumPy`. The two libraries work well together, and they meet different needs.

With `pandas`, it's always exciting to read a CSV file for the first time and plot a histogram. With `NumPy`, there's no real exciting moment. It's just useful. It's a powerful and versatile tool, and it will save you in a pinch, 9 times out of 10, when `pandas` doesn't do what you need to do.

Setup

As with `pandas`, if you are working from within a notebook environment, `NumPy` is probably already installed. However, if you need to follow the steps, you can follow the official installation guide: `https://numpy.org/install/`.

Starter functionality

If you want to play around with `NumPy` to get a feel for what it can do, please check out one of their quickstart tutorials, such as `https://numpy.org/doc/stable/user/quickstart.html`. In the tutorial, they explain some basic operations such as generating random numbers, reshaping arrays, printing arrays, and more.

Documentation

You can learn a lot more about `NumPy` at `https://numpy.org/doc/stable/`. In this book, we will rely on `NumPy` at a pinch, as I described previously, when `pandas` or another library is unable to do something that we need. Often, there is a simple solution in `NumPy`. So, keep an eye out for our use of `NumPy`, and learn what the process is doing.

Let's move on from analysis and processing to the libraries we'll use for visualizing our data.

Data visualization

There are several Python libraries that can be used for data visualizations. **Matplotlib** is a good place to start, but other libraries such as **Seaborn** can create more attractive visualizations, and **Plotly** can create interactive visualizations.

Matplotlib

Matplotlib is a Python library for data visualization. That's it. If you have data, Matplotlib can probably be used to visualize it. The library is integrated directly into pandas, so if you use pandas, you likely also use Matplotlib.

Matplotlib has a very steep learning curve and is not intuitive at all. I consider it a necessary evil if you are learning about Python data science. No matter how much data visualization I do with Matplotlib, it never becomes easy, and I memorize very little. I say all of this not to badmouth Matplotlib, but so that you will not feel negativity toward yourself if you struggle with the library. We all struggle with the library.

Setup

As with pandas and NumPy, if you are working from within a notebook environment, Matplotlib is probably already installed. However, if you need to follow the steps, you can follow the official installation guide: https://matplotlib.org/stable/users/installing/.

Starter functionality

If you use pandas to do data visualization, you are already using Matplotlib. For instance, in the pandas discussion, we used this code to plot a histogram:

```
df['petal width (cm)'].plot.hist()
```

Simple visualizations such as bar charts can easily be done. They're most easily done directly in pandas.

For instance, you could create a horizontal bar chart:

```
import matplotlib.pyplot as plt
 df['petal width (cm)'].value_counts().plot.
barh(figsize=(8,6)).invert_yaxis()
```

This would render a very simple data visualization.

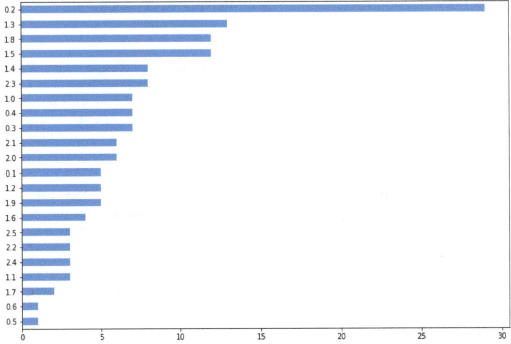

Figure 3.5 – Petal width horizontal bar chart

Documentation

The Matplotlib tutorials section is probably the best place to start learning: https://matplotlib.org/stable/tutorials/.

However, we will be making a lot of use of pandas and Matplotlib in this book, so you should pick up quite a bit of knowledge just from following along and practicing.

Seaborn

Seaborn is essentially "pretty Matplotlib." It is an extension of Matplotlib and produces more attractive data visualizations. The downside is that some things that were not remotely intuitive in Matplotlib are even less intuitive in Seaborn. It's got a learning curve like Matplotlib, and the learning never sticks. I find myself googling the web every time I need to make a visualization.

However, once you begin figuring things out, it's not so bad. You start to remember the headaches you overcame, and powering through difficult things that should be simple (such as resizing visualizations) becomes slightly less aggravating over time.

The visualizations are aesthetically pleasing, much more than the `Matplotlib` default. It's a lot of fun to just look through the `Seaborn` catalog of visualizations: `https://seaborn.pydata.org/examples/`.

We will occasionally use `Seaborn` in this book, but we will mostly use `Matplotlib` and `scikit-network` for our visualizations. Do learn about the library, though. It is important.

Setup

I don't remember whether notebooks come with `Seaborn` installed. It wouldn't surprise me if they do by now. However, if you need to install `Seaborn`, you can follow the official installation guide: `https://seaborn.pydata.org/installing.html`.

Starter functionality

`Seaborn` can do what `Matplotlib` can do, but better, and more prettily. For instance, let's redo the histogram we previously created, and include the **Kernel Density Estimate (KDE)**:

```
import seaborn as sns
sns.histplot(df['petal width (cm)'], kde=True)
```

This produces the following graph:

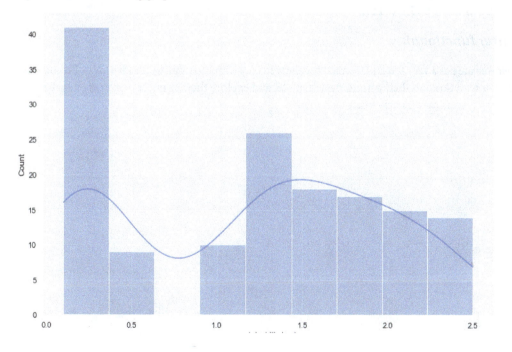

Figure 3.6 – Seaborn petal width histogram with the KDE

The best way to get started with Seaborn is to just jump in. Get some data, then try to visualize it. Look through their example gallery for something applicable to your data, and then try to build the same thing. Copy their code at first, and experiment with parameters.

Documentation

You can learn more about `Seaborn` from their website: `https://seaborn.pydata.org/`.

Plotly

To be honest, we will not use `Plotly` very often in this book, but there is one chapter where we needed to use an interactive scatterplot, and `Plotly` had a visualization that was useful. However, it is still good to know about `Plotly`, as it can be a powerful tool when you need simple intractability for data visualization.

`Plotly` advertises that it makes interactive, publication-quality graphs. That's it. It essentially does what `Matplotlib` and `Seaborn` do but in an interactive way. Building interaction into `Matplotlib` is possible, but it is painful. With `Plotly`, it is easy and fast, and a joy to do.

Setup

To install `Plotly`, follow their official guide: `https://plotly.com/python/getting-started/#installation`.

Starter functionality

Their website has a very fun and interactive gallery to play with, to stimulate creative ideas. I recommend that you read through their getting started guide, and explore their examples. `https://plotly.com/python/getting-started/`.

We will use `Plotly` in *Chapter 11, Unsupervised Machine Learning on Network Data.*

Figure 3.7 – Plotly interactive scatterplot

Documentation

You can learn a lot more about `Plotly` at their website: `https://plotly.com/python/`.

Now that we have covered some useful libraries for data visualization, let's do the same but for NLP and working with text.

NLP

Moving on from data visualization to NLP, there are also several Python libraries that will be helpful to us in processing text. Each library has unique offerings as well as strengths and weaknesses, and the documentation should be read.

Natural Language Toolkit

The **Natural Language Toolkit** (NLTK) is a Python library for working with natural language – language used by humans in our daily interactions and activities. NLTK is an older Python library, and it is often better to use other libraries such as spaCy for things that used to be done with NLTK, such as named-entity recognition or part-of-speech tagging.

However, just because it is older doesn't mean that it is obsolete. It is still very useful for analyzing text data, and it provides more linguistic capabilities than libraries such as spaCy.

Put simply, NLTK is a foundational library for doing NLP in Python, and you should not skip past it for more recent libraries and approaches.

Setup

Installing NLTK is easy. Just follow the guide: `https://www.nltk.org/install.html`.

Starter functionality

We're going to use NLTK quite a bit in this book. When you are ready, jump into the next chapter and we'll get started right away. Nothing really stands out as exciting to show off as starter functionality. NLP becomes more exciting the more you learn. It is a bit confusing in the beginning, to say the least.

Documentation

The NLTK website is not very helpful as a learning tool, but you can see some code snippets and examples here: `https://www.nltk.org`.

These days, you will likely learn more about NLTK from blog posts (*TowardsDataScience* and others) than from the official NLTK website. Do a Google search for `nltk getting started` and you'll find some helpful guides if you want to learn from other sources.

This book will be enough to help you get started with NLTK, but I do recommend learning from others who write about NLP. There is so much to learn and explore.

spaCy

spaCy is a more recent Python library for NLP, and some of its functionality can serve as a replacement for things that used to be done in NLTK. For instance, if you are doing named-entity recognition, part-of-speech tagging, or lemmatization, I recommend that you use spaCy, as it is quicker and simpler to get good results. However, the NLTK data loaders, stop words, tokenizers, and other linguistic offerings are still useful, and sometimes unmatched by spaCy. In my opinion, if you work with NLP in Python, you should learn about both NLTK and spaCy, not just spaCy. Learn about both, and remember where spaCy shines brighter than NLTK.

Setup

To install `spaCy`, please use the official installation guide: `https://spacy.io/usage#quickstart`.

Starter functionality

You're going to see `spaCy` used quite a bit in this book, especially for named-entity recognition, which is useful in graph construction. Feel free to skip ahead if you want to see it in use.

As an example, let's install the language model, like so:

```
!python -m spacy download en_core_web_sm
```

We can run this code:

```
import spacy
nlp = spacy.load("en_core_web_sm")
doc = nlp("David Knickerbocker wrote this book, and today is
September 4, 2022.")
for ent in doc.ents:
    print(ent.text, ent.label_)
```

And this would output some NER results:

```
David Knickerbocker PERSON
today DATE
September 4, 2022 DATE
```

How is it doing this, you might wonder? It's MAGIC! Just kidding – it's ML.

Documentation

Unlike `NLTK`, the `spaCy` documentation is thorough and helpful, and it looks modern. I recommend spending significant time on their website and learning about the capabilities. There is much more that can be done with `spaCy` than what we will cover in this book. To learn more, visit `https://spacy.io/usage/linguistic-features`.

Network analysis and visualization

Our next set of libraries is useful for analyzing and visualizing various types of networks. Network visualization is another form of data visualization but specifically for networks. It is very difficult and tedious to create useful network visualizations without network visualization software.

NetworkX

NetworkX is incredibly useful for doing social network analysis and network science using Python. Out of all of the Python libraries covered in this chapter, this library is the heart and soul of what you will be learning in this book. Everything revolves around `NetworkX`, and there is simply no better resource for doing network analysis in Python.

There are other libraries that can be used for very simple network visualization. In this book, I disregard all of those and show how `NetworkX` and `scikit-network` can be used to both do good network analysis and make reasonably attractive network visualizations.

There are also other libraries that are useful alongside `NetworkX`, for community detection, or for analyzing the fragility of networks, for instance, but `NetworkX` is the core. My goal for writing this book is to spread awareness of network science and using networks in software development, and `NetworkX` is crucial for this. I want more software engineers to learn these skills.

Setup

Please follow the official guide to installing `NetworkX`: `https://networkx.org/documentation/stable/install.html`.

Starter functionality

We're going to do a lot of really cool stuff with `NetworkX`, but if you want to check that it is working, you can run this code:

```
import networkx as nx
G = nx.les_miserables_graph()
sorted(G.nodes)[0:10]
```

In a notebook, this should show a Python list of the first 10 nodes included in the *Les Miserables* graph:

```
['Anzelma',
 'Babet',
 'Bahorel',
 'Bamatabois',
 'BaronessT',
 'Blacheville',
 'Bossuet',
 'Boulatruelle',
 'Brevet',
 'Brujon']
```

Or, you could do this, to get a quick summary of the *Les Miserables* graph:

```
print(nx.info(G))
Graph with 77 nodes and 254 edges
```

Documentation

The best way to learn about NetworkX is actually probably reading this book. There are very few books about using NetworkX. There are some blog posts, and a few of the guides on the NetworkX website can be helpful. There's this one, but it assumes that you have an understanding of graphs: https://networkx.org/documentation/stable/tutorial.html.

The website does have very detailed documentation on the different tools offered: https://networkx.org/documentation/stable/reference/.

My favorite part about their online resources is that they point to academic papers that were written about the functions that have been implemented. For instance, if you visit the page in the **Link Prediction** section about **Preferential Attachment**, you will see this block:

References

[1]   D. Liben-Nowell, J. Kleinberg. The Link Prediction Problem for Social Networks (2004). http://www.cs.cornell.edu/home/kleinber/link-pred.pdf

Figure 3.8 – NetworkX reference

On several occasions, I've enjoyed digging into the writing that has gone into the development of several algorithms. It is nice to know the background, and algorithms absolutely should not be used blindly. If you're going to use a tool, learn how it works, and where it doesn't work.

scikit-network

scikit-network advertises itself as a Python package for the analysis of large graphs, but this is not accurate at all. It is a Python package that excels at the visualization of small to medium-sized graphs but has little usefulness outside of that. However, for network visualization, I have found nothing better. NetworkX visualizations are slow, to the point that I will not teach them. They take a long time to render, even on a small network, and their appearance is bland and basic. scikit-network visualizations load very fast, on the other hand, as they are rendered SVGs. The visualizations load at a reasonable speed, even if you have hundreds or thousands of nodes.

However, the notion that this is a package for the analysis of large graphs is not accurate at all. This is visualization software, at its core, even if it is capable of doing a bit more than that. It seems they have added some additional capabilities (such as **Graph Neural Networks (GNNs)**), but most documentation pages highlight visualization. Do take a look at their offerings on embeddings and community detection, though. It seems that they are working toward doing more than network visualization and that the library is evolving.

Setup

To install `scikit-network`, please follow the official guide: `https://scikit-network.readthedocs.io/en/latest/first_steps.html`.

Starter functionality

In this book, we will only use `scikit-network` for network visualization, so if you'd like to get started with it, please take a look at some of their tutorials, such as `https://scikit-network.readthedocs.io/en/latest/tutorials/visualization/graphs.html`. There are several tutorials you can use to learn more about the library.

Documentation

The `scikit-network` documentation is pretty good and pretty well-organized. The simplicity of the documentation shows that it's really not a very deep Python library. If you want to learn more, start here: `https://scikit-network.readthedocs.io/en/latest/`.

You might be wondering why I only use this for visualization when it also offers capabilities for some ML, community detection, and other things. The reason is this: `NetworkX` and `KarateClub` also offer similar capabilities, and they do them better. For instance, `KarateClub` offers **Scalable Community Detection (SCD)** as an approach for **Graph Machine Community Detection**, and `NetworkX` provides the **Louvain method**. While `scikit-network` provides functionality for calculating **PageRank**, `NetworkX` does it better and more cleanly.

That is why I only show how it can be used for visualization. The other capabilities haven't shown to be better than those provided by `NetworkX` and `KarateClub`. It's nothing personal. We all have our favorite tools.

In the next section, we'll talk about the libraries we can use for ML.

ML

There is also a set of Python libraries for doing machine learning. Knowing the Python libraries and algorithms is not enough. Please learn as much as you can about the models and techniques you will use. There is always more to learn. Here are some of the libraries you will most likely frequently use.

scikit-learn

scikit-learn is a powerhouse in ML in Python. There are other important Python libraries for ML, such as `TensorFlow` and `PyTorch`, but we will not use them directly in this book. Just be aware that they are out there, and that you should learn about them as your ML skills improve.

If you are learning about or doing ML in Python, you will very quickly run into `scikit-learn`. `scikit-learn` provides a treasure trove of ML models, data loaders, preprocessors, transformations, and much more. There are many books and blog posts about `scikit-learn`, including several that have been published by Packt. If you are interested in learning ML, you will be doing a lot of reading, and `scikit-learn` is an excellent place to start.

Setup

If you are working with notebooks, then `scikit-learn` is probably already installed. However, if you need to install it, please follow the official guide: `https://scikit-learn.org/stable/install.html`.

Starter functionality

ML is complicated, and it takes several steps to build and use a model. For now, just make sure that `scikit-learn` is installed and that you can see which version you are running:

```
import sklearn
sklearn.__version__
```

For me, this shows that I am running version `1.0.2`. For the exercises in this book, we do not need to have an exact match, but for reproducibility in the real world, all versions of libraries should match when deploying anything.

Documentation

There are more resources for `scikit-learn` than there are for probably every other Python library I've mentioned in this chapter. ML is a trendy topic, and there is a demand for books, blog posts, videos, and other instruction, so there is no shortage of resources to learn from. However, the `scikit-learn` documentation is the place to start.

The `scikit-learn` starter guide can be found here: `https://scikit-learn.org/stable/getting_started.html`.

The full user guide can be found at `https://scikit-learn.org/stable/user_guide.html`, and there is also an API reference at `https://scikit-learn.org/stable/modules/classes.html`.

We will be doing a bit of supervised and unsupervised machine learning in this book, but we won't go very far into the topic. There will be much more to learn if you are interested. This book will only scratch the surface of ML and hopefully will kick off an obsession.

Karate Club

Karate Club advertises itself as being an unsupervised machine learning extension library for NetworkX. In other words, Karate Club bridges the gap between machine learning and graph analysis. It provides several useful unsupervised learning models that can be used for community detection as well as creating node embeddings from graphs that may be used in downstream supervised learning tasks. This is often referred to as graph machine learning, and there is a lot of excitement about using graphs and machine learning together, for instance, to predict new connections (who should be friends, who would like certain music or products, and so on).

What's the difference between this and scikit-learn? Well, scikit-learn can be used with any numeric data. Karate Club specializes in transforming graphs into numeric data through unsupervised machine learning. Karate Club outputs can be scikit-learn inputs, but rarely the other way around.

For Karate Club to be useful, you need graphs (NetworkX) and you need ML (scikit-learn, TensorFlow, PyTorch, and so on). Karate Club is not very useful by itself.

Karate Club is not yet supported by higher versions of Python, such as 3.10.

Setup

To install Karate Club, follow the official guide: https://karateclub.readthedocs.io/en/latest/notes/installation.html.

Starter functionality

There's no starter functionality worth showcasing. ML is complicated and takes several steps to be of use. For now, let's just verify that the library is installed and that we can see the version number:

```
import karateclub
karateclub.__version__
```

I can see that I am running version 1.3.0.

Documentation

Karate Club provides an introductory guide at https://karateclub.readthedocs.io/en/latest/notes/introduction.html. Here, you can start to see how the library may be used. You will also likely notice some references to other libraries, such as scikit-learn.

Karate Club has exemplary documentation. You can find it at `https://karateclub.readthedocs.io/en/latest/modules/root.html`. The thing that I love the most is that they cite the journal articles that were written about the models. For instance, if you look at the documentation for any of the algorithms, you will often find mention of the paper that was published about the model.

class **SCD**(*iterations: int = 25, eps: float = 1e-06, seed: int = 42*)   [source]

> An implementation of "SCD" from the WWW '14 paper "High Quality, Scalable and Parallel Community Detection for Large Real Graphs". The procedure greedily optimizes the approximate weighted community clustering metric. First, clusters are built around highly clustered nodes. Second, we refine the initial partition by using the approximate WCC. These refinements happen for the whole vertex set.

> Parameters:
> - **iterations** (*int*) – Refinemeent iterations. Default is 25.
> - **eps** (*float*) – Epsilon score for zero division correction. Default is 10**-6.
> - **seed** (*int*) – Random seed value. Default is 42.

Figure 3.9 – Karate club documentation reference

As I've looked through several of the Karate Club models for work in this book, I've had the opportunity to read many of these papers, and that has brought me many hours of enjoyment. I love this so much.

And if you are working with Karate Club in code, you can also see mention of the paper in notebooks if you *Shift + Tab* into a function call:

```
Init signature: SCD(iterations: int = 25, eps: float = 1e-06, seed: int = 42)
Docstring:
An implementation of `"SCD" <http://wwwconference.org/proceedings/www2014/proceedings/p225.pdf>`_ from the
WWW '14 paper "High Quality, Scalable and Parallel Community Detection for
Large Real Graphs". The procedure greedily optimizes the approximate weighted
community clustering metric. First, clusters are built around highly clustered nodes.
Second, we refine the initial partition by using the approximate WCC. These refinements
happen for the whole vertex set.

Args:
    iterations (int): Refinemeent iterations. Default is 25.
    eps (float): Epsilon score for zero division correction. Default is 10**-6.
    seed (int): Random seed value. Default is 42.
Init docstring: Creating an estimator.
```

Figure 3.10 – Karate club code reference

Not only is this library exciting to use but it's also educational to learn from.

spaCy (revisited)

Yes, I'm mentioning `spaCy` again. Why? Because `spaCy` offers several language models, and in many different languages. This means that you can use the `spaCy` pre-trained machine learning models for your own purposes, for instance, for named-entity recognition. Their models have already been trained and can be directly imported for use. You will learn how to do so in this book, in the next few chapters.

Summary

There are other Python libraries that we will use in this book, but they will be explained in their relevant chapters. In this chapter, I wanted to describe the primary libraries we will use for our work. In order to get to the really experimental stuff that we cover in this book, a foundation needs to be in place.

For instance, you need to be able to read and analyze tabular (structured) data. You also need to be able to visualize data. With text, you need to be able to convert text into a format that is ready for analysis and use. For graphs, you need to be able to do the same. And finally, if you want to apply machine learning to networks or text, you should understand how to do that.

That is why the sections have been broken down into data analysis and processing, data visualization, natural language processing, network analysis and visualization, and machine learning. I hope the structure helps.

With these libraries installed and briefly explained, we are now ready to get our hands dirty with network science, social network analysis, and NLP, all in Python.

Part 2: Graph Construction and Cleanup

Network data does not always exist for research that is interesting to an analyst. The chapters in this part show how to pull text data from various online resources and convert text data into networks that can be analyzed. These chapters also walk through data wrangling steps for cleaning networks for use in analysis and downstream processes.

This section includes the following chapters:

- *Chapter 4, NLP and Network Synergy*
- *Chapter 5, Even Easier Scraping*
- *Chapter 6, Graph Construction and Cleaning*

4

NLP and Network Synergy

In the previous chapters, we discussed **natural language processing** (**NLP**), network analysis, and the tools used in the Python programming language for both. We also discussed non-programmatic tools for doing network analysis.

In this chapter, we are going to put all of that knowledge to work. I hope to explain the power and insights that can be unveiled by combining NLP and network analysis, which is the theme of this book. In later chapters, we will continue with this theme, but we'll also discuss other tools for working with NLP and networks, such as unsupervised and supervised machine learning. This chapter will demonstrate techniques for determining who or what a piece of text is talking about.

The following topics will be covered in this chapter:

- Why are we learning about NLP in a network book?
- Asking questions to tell a story
- Introducing web scraping
- Choosing between libraries, APIs, and source data
- Using the Natural Language Toolkit (NLTK) library for part-of-speech (PoS) tagging
- Using spaCy for PoS tagging and named-entity recognition (NER)
- Converting entity lists into network data
- Converting network data into networks
- Doing a network visualization spot check

Technical requirements

In this chapter, we will be using several different Python libraries. The `pip install` command is listed in each section for installing each library, so just follow along and do the installations as needed. If you run into installation problems, there is usually an answer on Stack Overflow. Google the error!

Before we start, I would like to explain one thing so that the number of libraries we are using doesn't seem so overwhelming. It is the reason why we use each library that matters.

For most of this book, we will be doing one of three things: network analysis, network visualization, or using network data for machine learning (also known as GraphML).

Anytime we have network data, we will be using `NetworkX` to use it.

Anytime we are doing analysis, we will probably be using `pandas`.

The relationship looks like this:

- **Network**: `NetworkX`
- **Analysis**: `pandas`
- **Visualization**: `scikit-network`
- **ML**: `scikit-learn` and `Karate Club`

Look at the first words. If I want to do network analysis, then I'll be using `NetworkX` and `pandas`. If I want to do network visualization, I will be using `NetworkX` and `scikit-network`. If I want to do machine learning using network data, I will be using `NetworkX`, `scikit-learn`, and possibly `Karate Club`.

These are the core libraries that will be used in this chapter.

Also, *you must keep the code for the* `draw_graph()` *function handy*, as you will be using it throughout this book. That code is a bit gnarly because, at the time of writing, it needs to be. Unfortunately, `NetworkX` is not great for network visualization, and `scikit-network` is not great for network construction or analysis, so I use both libraries together for visualization, and that works well. I am hoping this will not be the case for later editions of this book and that visualization will be improved and simplified over time.

All of the necessary code can be found at `https://github.com/PacktPublishing/Network-Science-with-Python`.

Why are we learning about NLP in a network book?

I briefly answered this question in the introduction of the first chapter, but it is worth repeating in more detail. Many people who work on text analysis are aware of sentiment analysis and text classification. **Text classification** is the ability to predict whether a piece of text can be classified as something. For example, let's take this string:

"How are you doing today?"

What can we tell about this string? Is it a question or a statement? It's a question. Who is being asked this question? You are. Is there a positive, negative, or neutral emotion tied to the question? It looks neutral to me. Let's try another string.

"Jack and Jill went up the hill, but Jack fell down because he is an idiot."

This is a statement about Jack and Jill, and Jack is being called an idiot, which is not a very nice thing to say. However, it seems like the insult was written jokingly, so it is unclear whether the author was angry or laughing when it was written. I can confirm that I was laughing when I wrote it, so there is a positive emotion behind it, but text classification would struggle to pick up on that. Let's try another.

"I have never been so angry in my entire life as I am right now!"

The author is expressing a very strong negative sentiment. Sentiment analysis and text classification would easily pick up on that.

Sentiment analysis is a set of techniques that use algorithms to automatically detect emotions that are embedded into a piece of text or transcribed recording. Text classification uses the same algorithms, but the goal is not to identify an emotion but rather a theme. For instance, text classification can detect whether there is abusive language in text.

This is not a chapter about sentiment analysis or text classification. The purpose of this chapter is to explain how to automatically extract the entities (people, places, things, and more) that exist in text so that you can identify and study the social network that is being described in the text. However, text classification and sentiment analysis can be coupled with these techniques to provide additional context about social networks, or for building specialized social networks, such as friendship networks.

Asking questions to tell a story

I approach my work from a storytelling perspective; I let my story dictate the work, not the other way around. For instance, if I am starting a project, I'll ponder or even write down a series of who, what, where, when, why, and how questions:

- What data do we have? Is it enough?
- Where do we get more?
- How do we get more?

- What blockers prevent us from getting more?
- How often will we need more?

But this is a different kind of project. We want to gain deeper insights into a piece of text than we might gather through reading. Even after reading a whole book, most people can't memorize the relationships that are described in the text, and it would likely be a faulty recollection anyway. But we should have questions such as these:

- Who is mentioned in the text?
- Who do they know?
- Who are their adversaries?
- What are the themes of this text?
- What emotions are present?
- What places are mentioned?
- When did this take place?

It is important to come up with a solid set of questions before starting any kind of analysis. More questions will come to you the deeper you go.

This chapter will give you the tools to automatically investigate all of these questions other than the ones about themes and adversaries. The knowledge from this chapter, coupled with an understanding of text classification and sentiment analysis, will give you the ability to answer *all* these questions. That is the "why" of this chapter. We want to automatically extract people, places, and maybe even some things that are mentioned. Most of the time, I only want people and places, as I want to study the social network.

However, it is important to explain that this is useful for accompanying text classification and sentiment analysis. For example, a positive Amazon or Yelp review is of little use if you do not know what is being reviewed.

Before we can do any work to uncover the relationships that exist in text, we need to get some text. For practice, we have a few options. We could load it using a Python library such as NLTK, we could harvest it using the Twitter (or another social network) library, or we could scrape it ourselves. Even scraping has options, but I am only going to explain one way: using Python's `BeautifulSoup` library. Just know that there are options, but I love the flexibility of `BeautifulSoup`.

In this chapter, the demos will use text scraped off of the internet, and you can make slight changes to the code for your own web scraping needs.

Introducing web scraping

First, what even is **web scraping**, and who can do it? Anyone with any programming skill can do scraping using several different programming languages, but we will do this with Python. Web scraping is the action of harvesting content from web resources so that you may use the data in your products and software. You can use scraping to pull information that a website hasn't exposed as a data feed or through an API. But one warning: do not scrape too aggressively; otherwise, you could knock down a web server through an accidental **denial-of-service (DoS)** attack. Just get what you need as often as you need it. Go slow. Don't be greedy or selfish.

Introducing BeautifulSoup

BeautifulSoup is a powerful Python library for scraping anything that you have access to online. I frequently use this to harvest story URLs from news websites, and then I scrape each of these URLs for their text content. I typically do not want the actual HTML, CSS, or JavaScript, so I render the web page and then scrape the content.

`BeautifulSoup` is an important Python library to know about if you are going to be doing web scraping. There are other options for web scraping with Python, but `BeautifulSoup` is commonly used.

There is no better way to explain `BeautifulSoup` than to see it in action, so let's get to work!

Loading and scraping data with BeautifulSoup

In this hands-on demonstration, we will see three different ways of loading and scraping data. They are all useful independently, but they can also be used together in ways.

First, the easiest approach is to use a library that can get exactly what you want in one shot, with minimal cleanup. Several libraries such as `pandas`, `Wikipedia`, and `NLTK` have ways to load data, so let's start with them. As this book is primarily about extracting relationships from text and then analyzing them, we want text data. I will demonstrate a few approaches, and then describe the pros and cons of each.

Python library – Wikipedia

There is a powerful library for pulling data from Wikipedia, also called `Wikipedia`. It can be installed by running the following command:

```
pip install wikipedia
```

Once it has been installed, you may import it into your code like this:

```
import wikipedia as wiki
```

Once it has been imported, you have access to Wikipedia programmatically, allowing you to search for and use any content you are curious about. As we will be doing a lot of social network analysis while working through this book, let's see what `Wikipedia` has on the subject:

```
search_string = 'Social Network Analysis'
page = wiki.page(search_string)
content = page.content
content[0:680]
```

The last line, `content[0:680]`, just shows what is inside the `content` string, up to the 680th character, which is the end of the sentence shown in the following code. There is a lot more past 680. I chose to cut it off for this demonstration:

```
'Social network analysis (SNA) is the process of investigating
social structures  through the use of networks and graph
theory. It characterizes networked structures in terms of nodes
(individual actors, people, or things within the network) and
the ties, edges, or links (relationships or interactions)
that connect them.  Examples of social structures commonly
visualized through social network analysis include social media
networks, memes spread, information circulation, friendship
and acquaintance networks, business networks, knowledge
networks, difficult working relationships, social networks,
collaboration graphs, kinship, disease transmission, and sexual
relationships.'
```

With a few lines of code, we were able to pull text data directly from Wikipedia. Can we see what links exist on that Wikipedia page? Yes, we can!

```
links = page.links
links[0:10]
```

I'm using a square bracket, [, to select only the first 10 entries from `links`:

```
['Actor-network theory',
 'Adjacency list',
 'Adjacency matrix',
 'Adolescent cliques',
 'Agent-based model',
```

```
'Algorithm',
'Alpha centrality',
'Anatol Rapoport',
'Anthropology',
'ArXiv (identifier)']
```

If we are only in the A-links after choosing only the first 10 links, I think it's safe to say that there are probably many more. There's quite a lot to learn about social network analysis on Wikipedia! That was so simple! Let's move on to another option for easy scraping: the pandas library.

Python library – pandas

If you are interested in using Python for data science, you will inevitably learn about and use pandas. This library is powerful and versatile for working with data. For this demonstration, I will use it for pulling tabular data from web pages, but it can do much more. If you are interested in data science, learn as much as you can about pandas, and get comfortable using it.

pandas can be useful for pulling tabular data from web pages, but it is much less useful for raw text. If you want to load text from Wikipedia, you should use the Wikipedia library shown previously. If you want to pull text off other web pages, you should use the Requests and BeautifulSoup libraries together.

Let's use pandas to pull some tabular data from Wikipedia. First, let's try to scrape the same Wikipedia page using pandas and see what we get. If you have installed Jupyter on your computer, you likely already have pandas installed, so, let's jump straight to the code:

1. Start by importing pandas:

    ```
    import pandas as pd
    url = 'https://en.wikipedia.org/wiki/Social_network_
    analysis'
    data = pd.read_html(url)
    type(data)
    ```

    Here, we imported the pandas library and gave it a shorter name, and then used Pandas to read the same Wikipedia page on social network analysis. What did we get back? What does type(data) show? I would expect a pandas DataFrame.

2. Enter the following code. Just type data and run the code:

    ```
    data
    ```

You should see that we get a list back. In `pandas`, if you do a `read` operation, you will usually get a DataFrame back, so why did we get a list? This happened because there are multiple data tables on this page, so `pandas` has returned a Python list of all of the tables.

3. Let's check out the elements of the list:

    ```
    data[0]
    ```

    This should give us a `pandas` DataFrame of the first table from a Wikipedia page:

	0
0	Part of a series on
1	Sociology
2	History Outline Index
3	Methods Quantitative Qualitative Comparative C...
4	Subfields Criminology Culture Demography Devel...
5	PeopleEast Asia 1900s Fei Xiaotong South Asia ...
6	Perspectives Conflict theory Critical theory S...
7	Lists Bibliography Terminology Journals Organi...
8	Society portal WikiProject Sociology
9	.mw-parser-output .navbar{display:inline;font-...

Figure 4.1 – pandas DataFrame of the first element of Wikipedia data

This data is already looking a bit problematic. Why are we getting a bunch of text inside of a table? Why does the last row look like a bunch of code? What does the next table look like?

```
data[1]
```

This will give us a `pandas` DataFrame of the second table from the same Wikipedia page. Please be aware that Wikipedia pages are occasionally edited, so your results might be different:

	Network science
0	Theory
1	Graph Complex network Contagion Small-world Sc...
2	Network types
3	Informational (computing) Telecommunication Tr...
4	Graphs
5	Features Clique Component Cut Cycle Data struc...
6	Features
7	Clique Component Cut Cycle Data structure Edge...
8	Types
9	Bipartite Complete Directed Hyper Multi Random...
10	MetricsAlgorithms
11	Centrality Degree Motif Clustering Degree dist...
12	Models
13	Topology Random graph Erdős–Rényi Barabási–Alb...
14	Topology
15	Random graph Erdős–Rényi Barabási–Albert Fitne...
16	Dynamics
17	Boolean network agent based Epidemic/SIR
18	ListsCategories
19	Topics Software Network scientists Category:Ne...
20	vte

Figure 4.2 – pandas DataFrame of the second element of Wikipedia data

Gross. This looks even worse. I say that because it appears that we have some very short strings that look like web page sections. This doesn't look very useful. Remember, pandas is great for loading tabular data, but not great on tables of text data, which Wikipedia uses. We can already see that this is less useful than the results we very easily captured using the Wikipedia library.

4. Let's try a page that has useful data in a table. This page contains tabular data about crime in Oregon:

```
url = 'https://en.wikipedia.org/wiki/Crime_in_Oregon'
data = pd.read_html(url)
df = data[1]
df.tail()
```

This will show us the last five rows of a `Pandas` DataFrame containing Oregon crime data:

	Year	Population	Index	Violent	Property	Murder	Forcible rape	Robbery	Aggravated assault	Burglary	Larcenytheft	Vehicletheft
45	2005	3638871	170643	10444	160199	80	1266	2478	6620	27621	113316	19262
46	2006	3700758	145168	10373	135895	86	1195	2689	6403	23879	97556	14460
47	2007	3747455	142920	10777	132143	73	1255	2862	6587	22821	94773	14549
48	2008	3790060	134144	9747	124397	82	1156	2641	5868	20879	92187	11311
49	2009	3825657	123255	9744	113511	85	1168	2461	6030	19377	84265	9869

Figure 4.3 – Pandas DataFrame of tabular numeric Wikipedia data

Wow, this looks like useful data. However, the data does not show anything more recent than 2009, which is quite a while ago. Maybe there is a better dataset out there that we should use instead. Regardless, this shows that `Pandas` can easily pull tabular data off the web. However, there are a few things to keep in mind.

First, if you use scraped data in your projects, know that you are putting yourself at the mercy of the website administrators. If they decided to throw away the data table or rename or reorder the columns, that may break your application if it reads directly from Wikipedia. You can protect yourself against this by keeping a local copy of the data when you do a scrape, as well as including error handling in your code to watch for exceptions.

Be prepared for the worst. When you are scraping, build in any error-checking that you need as well as ways of knowing when your scrapers are no longer able to pull data. Let's move on to the next approach – using `NLTK`.

Python library – NLTK

Let's get straight to it:

1. First, NLTK does not come installed with Jupyter, so you will need to install it. You can do so with the following command:

    ```
    pip install nltk
    ```

2. Python's **NLTK** library can easily pull data from Project Gutenberg, which is a library of over 60,000 freely available books. Let's see what we can get:

    ```
    from nltk.corpus import gutenberg
    gutenberg.fileids()

    ...
    ['austen-emma.txt',
      'austen-persuasion.txt',
      'austen-sense.txt',
    ```

```
    'bible-kjv.txt',
    'blake-poems.txt',
    'bryant-stories.txt',
    'burgess-busterbrown.txt',
    'carroll-alice.txt',
    'chesterton-ball.txt',
    'chesterton-brown.txt',
    'chesterton-thursday.txt',
    'edgeworth-parents.txt',
    'melville-moby_dick.txt',
    'milton-paradise.txt',
    'shakespeare-caesar.txt',
    'shakespeare-hamlet.txt',
    'shakespeare-macbeth.txt',
    'whitman-leaves.txt']
```

As you can easily see, this is far fewer than 60,000 results, so we are limited in what we can get using this approach, but it is still useful data for practicing NLP.

3. Let's see what is inside the `blake-poems.txt` file:

```
file = 'blake-poems.txt'

data = gutenberg.raw(file)
data[0:600]

...

'[Poems by William Blake 1789]\n\n \nSONGS OF INNOCENCE
AND OF EXPERIENCE\nand THE BOOK of THEL\n\n\n SONGS
OF INNOCENCE\n \n \n INTRODUCTION\n \n Piping down the
valleys wild,\n   Piping songs of pleasant glee,\n On a
cloud I saw a child,\n   And he laughing said to me:\n
\n "Pipe a song about a Lamb!"\n   So I piped with
merry cheer.\n "Piper, pipe that song again;"\n   So I
piped: he wept to hear.\n \n "Drop thy pipe, thy happy
pipe;\n   Sing thy songs of happy cheer:!"\n So I sang
the same again,\n   While he wept with joy to hear.\n \n
"Piper, sit thee down and write\n   In a book, that all
may read."\n So he vanish\'d'
```

We can load the entire file. It is messy in that it contains line breaks and other formatting, but we will clean that out. What if we want one of the other books from the full library of 60,000 books that are not on this list? Are we out of luck? Browsing the website, I can see that I can read the plain text of one of my favorite books, Franz Kafka's *The Metamorphosis*, at https://www.gutenberg. org/files/5200/5200-0.txt. Let's try to get that data, but this time, we will use Python's Requests library.

Python library – Requests

The **Requests** library comes pre-installed with Python, so there should be nothing to do but import it. Requests is used to scrape raw text off the web, but it can do more. Please research the library to learn about its capabilities.

> **Important note**
>
> If you use this approach, please note that you should not do this aggressively. It is okay to load one book at a time like this, but if you attempt to download too many books at once, or aggressively crawl all books available on Project Gutenberg, you will very likely end up getting your IP address temporarily blocked.

For our demonstration, let's import the library and then scrape the raw text from Gutenberg's offering of *The Metamorphosis*:

```
import requests
url = 'https://www.gutenberg.org/files/5200/5200-0.txt'
data = requests.get(url).text
data
```

…

```
'ï»¿The Project Gutenberg eBook of Metamorphosis, by Franz
Kafka\r\n\r\nThis eBook is for the use of anyone anywhere in
the United States and\r\nmost other parts of the world at no
cost and with almost no restrictions\r\nwhatsoever. You may
copy it, give it away or re-use it under the terms\r\nof the
Project Gutenberg License included with this eBook or online
at\r\nwww.gutenberg.org. If you are not located in the United
States, you\r\nwill have to check the laws of the country
where you are located before\r\nusing this eBook.\r\n\r\n**
This is a COPYRIGHTED Project Gutenberg eBook, Details Below
**\r\n**     Please follow the copyright guidelines in this
file.     *\r\n\r\nTitle: Metamorphosis\r\n\r\nAuthor: Franz
Kafka\r\n\r\nTranslator: David Wyllie\r\n\r\nRelease Date:
May 13, 2002 [eBook #5200]\r\n[Most recently updated: May 20,
2012]\r\n\r\nLanguage: English\r\n\r\nCharacter set encoding:
```

```
UTF-8\r\n\r\nCopyright (C) 2002 by David Wyllie.\r\n\r\n***
START OF THE PROJECT GUTENBERG EBOOK METAMORPHOSIS ***\r\n\
r\n\r\n\r\n\r\nMetamorphosis\r\n\r\nby Franz Kafka\r\n\r\
nTranslated by David Wyllie\r\n\r\n\r\n\r\n\r\nI\r\n\r\n\r\
nOne morning, when Gregor Samsa woke from troubled dreams,
he found\r\nhimself transformed in his bed into a horrible
vermin...'
```

I chopped off all text after "vermin" just to briefly show what is in the data. Just like that, we've got the full text of the book. As was the case with NLTK, the data is full of formatting and other characters, so this data needs to be cleaned before it will be useful. Cleaning is a very important part of everything that I will explain in this book. Let's keep moving; I will show some ways to clean text during these demonstrations.

I've shown how easy it is to pull text from Gutenberg or Wikipedia. But these two do not even begin to scratch the surface of what is available for scraping on the internet. pandas can read tabular data from any web page, but that is limited. Most content on the web is not perfectly formatted or clean. What if we want to set up scrapers to harvest text and content from various news websites that we are interested in? NLTK won't be able to help us get that data, and Pandas will be limited in what it can return. What are our options? We saw that the Requests library was able to pull another Gutenberg book that wasn't available through NLTK. Can we similarly use requests to pull HTML from websites? Let's try getting some news from Okinawa, Japan!

```
url = 'http://english.ryukyushimpo.jp/'
data = requests.get(url).text
data
'<!DOCTYPE html PUBLIC "-//W3C//DTD XHTML 1.0 Transitional//EN"
"http://www.w3.org/TR/xhtml1/DTD/xhtml1-transitional.dtd">\r\
n<html xmlns="http://www.w3.org/1999/xhtml" dir="ltr" lang="en-
US">\r\n\r\n<!-- BEGIN html head -->\r\n<head profile="http://
gmpg.org/xfn/11">\r\n<meta http-equiv="Content-Type"
content="text/html; charset=UTF-8" />\r\n<title>Ryukyu Shimpo -
Okinawa, Japanese newspaper, local news</title>...'
```

There we go! We've just loaded the raw HTML from the given URL, and we could do that for any publicly accessible web page. However, if you thought the Gutenberg data was messy, look at this! Do we have any hope in the world of using this, let alone building automation to parse HTML and mine useful data? Amazingly, the answer is yes, and we can thank the BeautifulSoup Python library and other scraping libraries for that. They have truly opened up a world of data for us. Let's see what we can get out of this using BeautifulSoup.

Python library – BeautifulSoup

First, BeautifulSoup is used along with the Requests library. Requests comes pre-installed with Python, but BeautifulSoup does not, so you will need to install it. You can do so with the following command:

```
pip install beautifulsoup4
```

There's a lot you can do with BeautifulSoup, so please go explore the library. But what would it take to extract all of the links from the Okinawa News URL? This is how you can do exactly that:

```
from bs4 import BeautifulSoup
soup = BeautifulSoup(data, 'html.parser')
links = soup.find_all('a', href=True)
links
[<a href="http://english.ryukyushimpo.jp">Home</a>,
 <a href="http://english.ryukyushimpo.jp">Ryukyu Shimpo -
Okinawa, Japanese newspaper, local news</a>,
 <a href="http://english.ryukyushimpo.jp/special-feature-
okinawa-holds-mass-protest-rally-against-us-base/">Special
Feature: Okinawa holds mass protest rally against US base</a>,
 <a href="http://english.ryukyushimpo.
jp/2021/09/03/34020/">Hirokazu Ueyonabaru returns home to
Okinawa from the Tokyo Paralympics with two bronze medals in
wheelchair T52 races, "the cheers gave me power"</a>,
 <a href="http://english.ryukyushimpo.
jp/2021/09/03/34020/"><img alt="Hirokazu Ueyonabaru returns
home to Okinawa from the Tokyo Paralympics with two bronze
medals in wheelchair T52 races, "the cheers gave me power""
class="medium" src="http://english.ryukyushimpo.jp/wp-content/
uploads/2021/09/RS20210830G01268010100.jpg"/> </a>…]
```

That was easy! I am only showing the first several extracted links. The first line imported the library, the second line set BeautifulSoup up for parsing HTML, the third line looked for all links containing an href attribute, and then finally the last line displayed the links. How many links did we harvest?

```
len(links)
…
277
```

Your result might be different, as the page may have been edited after this book was written.

277 links were harvested in less than a second! Let's see whether we can clean these up and just extract the URLs. Let's not worry about the link text. We should also convert this into a list of URLs rather than a list of <a> HTML tags:

```
urls = [link.get('href') for link in links]
urls
...
['http://english.ryukyushimpo.jp',
 'http://english.ryukyushimpo.jp',
 'http://english.ryukyushimpo.jp/special-feature-okinawa-holds-
mass-protest-rally-against-us-base/',
 'http://english.ryukyushimpo.jp/2021/09/03/34020/',
 'http://english.ryukyushimpo.jp/2021/09/03/34020/',
 'http://english.ryukyushimpo.jp/2021/09/03/34020/',
 'http://english.ryukyushimpo.jp/2021/09/03/34020/'...]
```

Here, we are using list comprehension and BeautifulSoup to extract the value stored in href for each of our harvested links. I can see that there is some duplication, so we should remove that before eventually storing the results. Let's see whether we lost any of the former 277 links:

```
len(urls)
...
277
```

Perfect! Let's take one of them and see whether we can extract the raw text from the page, with all HTML removed. Let's try this URL that I have hand-selected:

```
url = 'http://english.ryukyushimpo.jp/2021/09/03/34020/'
data = requests.get(url).text
soup = BeautifulSoup(data, 'html.parser')
soup.get_text()
...
"\n\n\n\n\nRyukyu Shimpo - Okinawa, Japanese newspaper, local
news  » Hirokazu Ueyonabaru returns home to Okinawa from the
Tokyo Paralympics with two bronze medals in wheelchair T52
races, "the cheers gave me power"\n\n\n\n\n\n\n\n\n\n\n\n\n\
n\n\n\n\n\n\n\n\n\n\n\n\n\n\n\n\n\n\n\n\n\n\n\n\n\n\n\n\n\n\
n\n\n\n\n\n\n\n\n\n\n\n\n\n\n\nHome\n\n\n\nSearch\n\n\n\n\
n\n\n\n\n\n\nTuesdaySeptember 07,2021Ryukyu Shimpo - Okinawa,
Japanese newspaper, local news\n\n\n\n\n\n\r\nTOPICS:Special
Feature: Okinawa holds mass protest rally against US base\n\
```

```
n\n\n\n\n\n\n\n\n\n\n\nHirokazu Ueyonabaru returns home to
Okinawa from the Tokyo Paralympics with two bronze medals in
wheelchair T52 races, "the cheers gave me power"…"
```

Done! We have captured pretty clean and usable text from a web page! This can be automated to constantly harvest links and text from any website of interest. Now, we have what we need to get to the fun stuff in this chapter: extracting entities from text, and then using entities to build social networks. It should be obvious by now that some cleanup will be needed for any of the options that we have explored, so let's just make short work of that now. To get this perfect, you need to do more than I am about to do, but let's at least make this somewhat usable:

```
text = soup.get_text()
text[0:500]

...

'\n\n\n\n\nRyukyu Shimpo – Okinawa, Japanese newspaper, local
news  » Hirokazu Ueyonabaru returns home to Okinawa from the
Tokyo Paralympics with two bronze medals in wheelchair T52
races, "the cheers gave me power"\n\n\n\n\n\n\n\n\n\n\n\n\n\
n\n\n\n\n\n\n\n\n\n\n\n\n\n\n\n\n\n\n\n\n\n\n\n\n\n\n\n\n\n\
n\n\n\n\n\n\n\n\n\n\n\n\n\nHome\n\n\n\nSearch\n\n\n\n\
n\n\n\n\n\nTuesdaySeptember 07,2021Ryukyu Shimpo – Okinawa,
Japanese newspaper, local news\n\n\n\n\n\r\nTOPICS:Special
Feature: Okinawa holds mass protest rally against US base\n\n\
n\n\n\n\n\n\n\n\nHirokazu Ueyonabaru returns home t'
```

The first thing that stands out to me is the amount of text formatting and special characters that exist in this text. We have a few options. First, we could convert all of the line spaces into empty spaces. Let's see what that looks like:

```
text = text.replace('\n', ' ').replace('\r', ' ').replace('\t',
' ').replace('\xa0', ' ')
text

...

"     Ryukyu Shimpo – Okinawa, Japanese newspaper, local
news  » Hirokazu Ueyonabaru returns home to Okinawa from the
Tokyo Paralympics with two bronze medals in wheelchair T52
races, "the cheers gave me power"
                              Home    Search
TuesdaySeptember 07,2021Ryukyu Shimpo – Okinawa, Japanese
newspaper, local news        TOPICS:Special Feature: Okinawa
holds mass protest rally against US base            Hirokazu
Ueyonabaru returns home to Okinawa from the Tokyo Paralympics
with two bronze medals in wheelchair T52 races, "the cheers
```

```
gave me power"    Tokyo Paralympic bronze medalist Hirokazu
Ueyonabaru receiving congratulations from some young supporters
at Naha Airport on August 30..."
```

If you scroll further down the text, you may see that the story ends at "Go to Japanese," so let's remove that as well as everything after:

```
cutoff = text.index('Go to Japanese')
cutoff
...
1984
```

This shows that the cutoff string starts at the 1,984[th] character. Let's keep everything up to the cutoff:

```
text = text[0:cutoff]
```

This has successfully removed the footer junk, but there is still some header junk to deal with, so let's see whether we can remove that. This part is always tricky, and every website is unique in some way, but let's remove everything before the story as an exercise. Looking closer, I can see that the story starts at the second occurrence of "Hirokazu Uevonabaru." Let's capture everything from that point and beyond. We will be using .rindex() instead of .index() to capture the last occurrence. This code is too specific for real-world use, but hopefully, you can see that you have options for cleaning dirty data:

```
cutoff = text.rindex('Hirokazu Ueyonabaru')
text = text[cutoff:]
```

If you are not familiar with Python, that might look a bit strange to you. We are keeping everything after the beginning of the last occurrence of "Hirokazu Ueyonabaru," which is where the story starts. How does it look now?

```
text
...
'Hirokazu Ueyonabaru, 50, - SMBC Nikko Securities Inc. - who
won the bronze medal in both the 400-meter and 1,500-meter
men's T52 wheelchair race, returned to Okinawa the evening of
August 30, landing at Naha airport. Seeing the people who came
out to meet him, he said "It was a sigh of relief (to win a
medal)" beaming a big smile. That morning he contended with
press conferences in Tokyo before heading home. He showed off
his two bronze medals, holding them up from his neck in the
airport lobby, saying "I could not have done it without all the
cheers from everyone..."'
```

That looks just about perfect! There is always more cleaning that can be done, but this is good enough! When you first get started with any new scraping, you will need to clean, inspect, clean, inspect, clean, and inspect – gradually, you will stop finding obvious things to remove. You don't want to cut too much. Just get the text to be usable – we will do additional cleaning at later steps.

Choosing between libraries, APIs, and source data

As part of this demonstration, I showed several ways to pull useful data off of the internet. I showed that several libraries have ways to load data directly but that there are limitations to what they have available. NLTK only offered a small portion of the complete Gutenberg book archive, so we had to use the `Requests` library to load *The Metamorphosis*. I also demonstrated that `Requests` accompanied by `BeautifulSoup` can easily harvest links and raw text.

Python libraries can also make loading data very easy when those libraries have data loading functionality as part of their library, but you are limited by what those libraries make available. If you just want some data to play with, with minimal cleanup, this may be ideal, but there will still be cleanup. You will not get away from that when working with text.

Other web resources expose their own APIs, which makes it pretty simple to load data after sending a request to them. Twitter does this. You authenticate using your API key, and then you can pull whatever data you want. This is a happy middle-ground between Python libraries and web scraping.

Finally, web scraping opens up the entire web to you. If you can access a web page, you can scrape it and use any text and data that it has made available. You have flexibility with web scraping, but it is more difficult, and the results require more cleanup.

I tend to approach my own scraping and data enrichment projects by making considerations in the following order:

- Is there a Python library that will make it easy for me to load the data I want?
- No? OK, is there an API that I can use to pull the data that I want?
- No? OK, can I just scrape it using `BeautifuSoup`? Yes? Game on. Let's dance.

Start simple and scale out in terms of complexity only as needed. Starting simple means starting with the simplest approach – in this case, Python libraries. If libraries won't help, add a little complexity by seeing whether an API is available to help and whether it is affordable. If one is not available, then web scraping is the solution that you need, and there is no way around it, but you will get the data that you need.

Now that we have text, we are going to move into NLP. Specifically, we will be using **PoS tagging and NER** as two distinct ways to extract entities (people and things) from raw text.

Using NLTK for PoS tagging

In this section, I will explain how to do what is called PoS tagging using the NLTK Python library. NLTK is an older Python NLP library, but it is still very useful. There are also pros and cons when comparing NLTK with other Python NLP libraries, such as spaCy, so it doesn't hurt to understand the pros and cons of each. However, during my coding for this demonstration, I realized just how much easier spaCy has made both PoS tagging as well as NER, so, if you want the easiest approach, feel free to just skip ahead to spaCy. I am still fond of NLTK, and in some ways, the library still feels more natural to me than spaCy, but that may simply be due to years of use. Anyway, I'd like to demonstrate PoS tagging with NLTK, and then I will demonstrate both PoS tagging and NER with spaCy in the next section.

PoS tagging is a process that takes text tokens and identifies the PoS that the token belongs to. Just as a review, a token is a single word. A token might be *apples*.

With NLP, tokens are useful, but bigrams are often even more useful, and they can improve the results for text classification, sentiment analysis, and even unsupervised learning. A bigram is essentially two tokens – for example, *two tokens*.

Let's not overthink this. It's just two tokens. What do you think a trigram is? That's right – three tokens. For instance, if filler words were removed from some text before the trigrams were captured, you could have one for *green eggs ham*.

There are many different pos_tags, and you can see the list here: https://www.ling.upenn. edu/courses/Fall_2003/ling001/penn_treebank_pos.html.

For the work that we are doing, we will only use the NLP features that we need, and PoS tagging and NER are two different approaches that are useful for identifying entities (people and things) that are being described in text. In the mentioned list, the ones that we want are NNP and NNPS, and in most cases, we'll find NNP, not NNPS.

To explain what we are trying to do, we are going to follow these steps:

1. Get some text to work with.
2. Split the text into sentences.
3. Split each sentence into tokens.
4. Identify the PoS tag for each token.
5. Extract each token that is a proper noun.

A proper noun is the name of a person, place, or thing. I have been saying that we want to extract entities and define entities as people, places, or things, so the NNP tag will identify exactly what we want:

1. Let's get to work and get some text data!

```
url = 'https://www.gutenberg.org/files/5200/5200-0.txt'
text = requests.get(url).text
```

We used this code previously to load the entire text from Kafka's book *The Metamorphosis*.

2. There is a lot of junk in the header of this file, but the story starts at "One morning," so let's remove everything from before that. Feel free to explore the `text` variable as we go. I am leaving out repeatedly showing the data to save space:

```
cutoff = text.index('One morning')
text = text[cutoff:]
```

Here, we have identified the starting point of the phrase, `One morning`, and then removed everything up to that point. It's all just header junk that we don't need.

3. Next, if you look at the bottom of the text, you can see that the story ends at, `*** END OF THE PROJECT GUTENBERG EBOOK METAMORPHOSIS`, so let's cut from that point onward as well:

```
cutoff = text.rindex('*** END OF THE PROJECT GUTENBERG
EBOOK METAMORPHOSIS ***')
text = text[:cutoff]
```

Look closely at the cutoff and you will see that the cutoff is in a different position from that used for removing the header. I am essentially saying, "*Give me everything up to the cutoff*." How does the ending text look now?

```
text[-500:]

...

'talking, Mr. and Mrs.\r\nSamsa were struck, almost
simultaneously, with the thought of how their\r\ndaughter
was blossoming into a well built and beautiful young
lady.\r\nThey became quieter. Just from each otherâ\
x80\x99s glance and almost without\r\nknowing it they
agreed that it would soon be time to find a good man\r\
nfor her. And, as if in confirmation of their new dreams
and good\r\nintentions, as soon as they reached their
destination Grete was the\r\nfirst to get up and stretch
out her young body.\r\n\r\n\r\n\r\n\r\n'
```

We have successfully removed both header and footer junk. I can see that there are a lot of line breaks, so let's

```
remove all of those as well.
text = text.replace('\r', ' ').replace('\n', ' ')
text
...
'One morning, when Gregor Samsa woke from troubled
dreams, he found  himself transformed in his bed into
a horrible vermin. He lay on his armour-like back, and
if he lifted his head a little he could see his  brown
belly, slightly domed and divided by arches into stiff
sections...'
```

Not bad. That's a noticeable improvement, and we are getting closer to clean text. Apostrophes are also being mangled, being shown as â\x80\x99, so let's replace those:

```
text = text.replace('â\x80\x99', '\'').replace('â\x80\
x9c', '"').replace('â\x80\x9d', '""')\
.replace('â\x80\x94', ' ')
print(text)
...
One morning, when Gregor Samsa woke from troubled dreams,
he found  himself transformed in his bed into a horrible
vermin. He lay on his  armour-like back, and if he lifted
his head a little he could see his  brown belly, slightly
domed and divided by arches into stiff sections.  The
bedding was hardly able to cover it and seemed ready to
slide off  any moment. His many legs, pitifully thin
compared with the size of the  rest of him, waved about
helplessly as he looked.    "What's happened to me?"" he
thought...
```

4. That is about perfect, so let's convert these steps into a reusable function:

```
def get_data():

    url = 'https://www.gutenberg.org/files/5200/5200-0.
txt'

    text = requests.get(url).text

    # strip header junk
    cutoff = text.index('One morning')
    text = text[cutoff:]
```

```
# strip footer junk
cutoff = text.rindex('*** END OF THE PROJECT
GUTENBERG EBOOK METAMORPHOSIS ***')
text = text[:cutoff]

# pre-processing to clean the text
text = text.replace('\r', ' ').replace('\n', ' ')
text = text.replace('â\x80\x99', '\'').replace('â\
x80\x9c', '"')\
    .replace('â\x80\x9d', '"').replace('â\x80\x94', '
')


return text
```

5. We should have fairly clean text after running this function:

```
text = get_data()
text

...

'One morning, when Gregor Samsa woke from troubled
dreams, he found  himself transformed in his bed into
a horrible vermin. He lay on his  armour-like back, and
if he lifted his head a little he could see his  brown
belly, slightly domed and divided by arches into stiff
sections.  The bedding was hardly able to cover it and
seemed ready to slide off  any moment. His many legs,
pitifully thin compared with the size of the  rest of
him, waved about helplessly as he looked.   "What\'s
happened to me?"" he thought'
```

Outstanding! We are now ready for the next steps.

Before we move on, I want to explain one thing: if you do PoS tagging on the complete text of any book or article, then the text will be treated as one massive piece of text, and you lose your opportunity to understand how entities relate and interact. All you will end up with is a giant entity list, which isn't very helpful for our needs, but it can be useful if you just want to extract entities from a given piece of text.

For our purposes, the first thing you need to do is split the text into sentences, chapters, or some other desirable bucket. For simplicity, let's do this by sentences. This can easily be done using NLTK's sentence tokenizer:

```
from nltk.tokenize import sent_tokenize

sentences = sent_tokenize(text)
sentences[0:5]
...

['One morning, when Gregor Samsa woke from troubled dreams, he
found  himself transformed in his bed into a horrible vermin.',
 'He lay on his  armour-like back, and if he lifted his head
a little he could see his  brown belly, slightly domed and
divided by arches into stiff sections.',
 'The bedding was hardly able to cover it and seemed ready to
slide off  any moment.',
 'His many legs, pitifully thin compared with the size of
the  rest of him, waved about helplessly as he looked.',
 '"What\'s happened to me?""']
```

Beautiful! We have a list of sentences to work with. Next, we want to take each of these sentences and extract any mentioned entities. We want the NNP-tagged tokens. This part takes a little work, so I will walk you through it. If we just feed the sentences to NLTK's pos_tag tool, it will misclassify everything:

```
import nltk
nltk.pos_tag(sentences)
[('One morning, when Gregor Samsa woke from troubled dreams, he
found  himself transformed in his bed into a horrible vermin.',
  'NNP'),
 ('He lay on his  armour-like back, and if he lifted his head
a little he could see his  brown belly, slightly domed and
divided by arches into stiff sections.',
  'NNP'),
 ('The bedding was hardly able to cover it and seemed ready to
slide off  any moment.',
  'NNP'),
 ('His many legs, pitifully thin compared with the size of
the  rest of him, waved about helplessly as he looked.',
  'NNP'),
```

```
('"What\'s happened to me?""', 'NNP'),
('he thought.', 'NN')…]
```

Good effort, but that is not what we need. What we need to do is go through each sentence and identify the PoS tags, so let's do this manually for a single sentence:

```
sentence = sentences[0]
sentence
```

…

```
'One morning, when Gregor Samsa woke from troubled dreams, he
found  himself transformed in his bed into a horrible vermin.'
```

First, we need to tokenize the sentence. There are many different tokenizers in NLTK, with strengths and weaknesses. I have gotten comfortable with the casual tokenizer, so I'll just use that. The casual tokenizer does well with casual text, but there are several other tokenizers available to choose from:

```
from nltk.tokenize import casual_tokenize
tokens = casual_tokenize(sentence)
tokens
```

…

```
['One',
 'morning',
 ',',
 'when',
 'Gregor',
 'Samsa',
 'woke',
 'from',
 'troubled',
 'dreams',
 ',',
 'he',
 'found',
 'himself',
 'transformed',
 'in',
 'his',
 'bed',
 'into',
```

```
'a',
'horrible',
'vermin',
'.']
```

Great. Now, for each token, we can find its corresponding pos_tag:

```
nltk.pos_tag(tokens)
...
[('One', 'CD'),
 ('morning', 'NN'),
 (',', ','),
 ('when', 'WRB'),
 ('Gregor', 'NNP'),
 ('Samsa', 'NNP'),
 ('woke', 'VBD'),
 ('from', 'IN'),
 ('troubled', 'JJ'),
 ('dreams', 'NNS'),
 (',', ','),
 ('he', 'PRP'),
 ('found', 'VBD'),
 ('himself', 'PRP'),
 ('transformed', 'VBN'),
 ('in', 'IN'),
 ('his', 'PRP$'),
 ('bed', 'NN'),
 ('into', 'IN'),
 ('a', 'DT'),
 ('horrible', 'JJ'),
 ('vermin', 'NN'),
 ('.', '.')]
```

That's also perfect! We are interested in extracting the NNPs. Can you see the two tokens that we want to extract? That's right, it's Gregor Samsa. Let's loop through these PoS tags and extract the NNP tokens:

```
entities = []
for row in nltk.pos_tag(tokens):
```

```
        token = row[0]
        tag = row[1]
        if tag == 'NNP':
            entities.append(token)
    entities

    ...

    ['Gregor', 'Samsa']
```

This is what we need. NER would hopefully identify these two results as one person, but once this is thrown into a graph, it's very easy to correct. Let's convert this into a function that will take a sentence and return the NNP tokens – the entities:

```
def extract_entities(sentence):
    entities = []
    tokens = casual_tokenize(sentence)
    for row in nltk.pos_tag(tokens):
        token = row[0]
        tag = row[1]
        if tag == 'NNP':
            entities.append(token)
    return entities
```

That looks good. Let's give it a shot!

```
extract_entities(sentence)

...

['Gregor', 'Samsa']
```

Now, let's be bold and try this out against every sentence in the entire book:

```
entities = [extract_entities(sentence) for sentence in
sentences]
entities
[['Gregor', 'Samsa'],
 [],
 [],
 [],
 ["What's"],
 [],
```

```
    [],
    [],
    ['Samsa'],
    [],
    ['Gregor'],
    [],
    [],
    [],
    [],
    ['Oh', 'God', '"', '"', "I've"],
    [],
    [],
    ['Hell']]
```

Just to make analysis a bit easier, let's do two things: first, let's replace those empty lists with None. Second, let's throw all of this into a Pandas DataFrame:

```
def extract_entities(sentence):
    entities = []
    tokens = casual_tokenize(sentence)
    for row in nltk.pos_tag(tokens):
        token = row[0]
        tag = row[1]
        if tag == 'NNP':
            entities.append(token)
    if len(entities) > 0:
        return entities
    else:
        return None
entities = [extract_entities(sentence) for sentence in
sentences]
entities
[['Gregor', 'Samsa'],
 None,
 None,
 None,
 ["What's"],
```

```
   None,
   None,
   None,
   ['Samsa'],
   None,
   ['Gregor'],
   None,
   None,
   None,
   None,
   ['Oh', 'God', '"', '"', "I've"],
   None,
   None,
   ['Hell']]
```

```
import pandas as pd
```

```
df = pd.DataFrame({'sentence':sentences, 'entities':entities})
df.head(10)
```

This will give us a DataFrame of sentences and entities extracted from sentences:

	sentence	entities
0	One morning, when Gregor Samsa woke from troub...	[Gregor, Samsa]
1	He lay on his armour-like back, and if he lif...	None
2	The bedding was hardly able to cover it and se...	None
3	His many legs, pitifully thin compared with th...	None
4	"What's happened to me?""	[What's]
5	he thought.	None
6	It wasn't a dream.	None
7	His room, a proper human room although a litt...	None
8	A collection of textile samples lay spread out...	[Samsa]
9	It showed a lady fitted out with a fur hat an...	None

Figure 4.4 – Pandas DataFrame of sentence entities

That's a good start. We can see that "**What's**" has been incorrectly flagged by NLTK, but it's normal for junk to get through when dealing with text. That'll get cleaned up soon. For now, we want to be able to build a social network using this book, so let's grab every entity list that contains two or more entities. We need at least two to identify a relationship:

```
df = df.dropna()
df = df[df['entities'].apply(len) > 1]
entities = df['entities'].to_list()
entities
```

```
[['Gregor', 'Samsa'],
 ['Oh', 'God', '"', '"', "I've"],
 ['"', '"'],
 ["I'd", "I'd"],
 ["I've", "I'll"],
 ['First', "I've"],
 ['God', 'Heaven'],
 ['Gregor', '"', '"', '"'],
 ['Gregor', "I'm"],
 ['Gregor', 'Gregor', '"', '"', '"'],
 ['Gregor', "I'm"],
 ['"', '"', 'Gregor', '"'],
 ['Seven', '"'],
 ["That'll", '"', '"'],
 ["They're", '"', '"', 'Gregor'],
 ["Gregor's", 'Gregor'],
 ['Yes', '"', 'Gregor'],
 ['Gregor', '"', '"'],
 ['Mr', 'Samsa', '"', '"']]
```

This looks pretty good other than some punctuation sneaking in. Let's revisit the previous code and disregard any non-alphabetical characters. That way, `Gregor's` will become `Gregor`, `I'd` will become `I`, and so forth. That'll make cleanup a lot easier:

```
def extract_entities(sentence):
    entities = []
    tokens = casual_tokenize(sentence)
    for row in nltk.pos_tag(tokens):
```

```
            token = row[0]
            tag = row[1]
            if tag == 'NNP':
                if "'" in token:
                    cutoff = token.index('\'')
                    token = token[:cutoff]
                entities.append(token)
    if len(entities) > 0:
        return entities
    else:
        return None
entities = [extract_entities(sentence) for sentence in
sentences]
entities
[['Gregor', 'Samsa'],
 None,
 None,
 None,
 ['What'],
 None,
 None,
 None,
 ['Samsa']…]
```

Let's put this back into a DataFrame and repeat our steps to see whether the entity list is looking better:

```
df = pd.DataFrame({'sentence':sentences, 'entities':entities})
df = df.dropna()
df = df[df['entities'].apply(len) > 1]
entities = df['entities'].to_list()
entities
[['Gregor', 'Samsa'],
 ['Oh', 'God', '"', '"', 'I'],
 ['"', '"'],
 ['I', 'I'],
 ['I', 'I'],
 ['First', 'I'],
```

```
['God', 'Heaven'],
['Gregor', '"', '"', '"'],
['Gregor', 'I'],
['Gregor', 'Gregor', '"', '"', '"'],
['Gregor', 'I'],
['"', '"', 'Gregor', '"'],
['Seven', '"'],
['That', '"', '"'],
['They', '"', '"', 'Gregor'],
['Gregor', 'Gregor'],
['Yes', '"', 'Gregor'],
['Gregor', '"', '"'],
['Mr', 'Samsa', '"', '"']]
```

This is getting better, but some double quotes are still in the data. Let's just remove any punctuation and anything after:

```
from string import punctuation
def extract_entities(sentence):
    entities = []
    tokens = casual_tokenize(sentence)
    for row in nltk.pos_tag(tokens):
        token = row[0]
        tag = row[1]
        if tag == 'NNP':
            for p in punctuation:
                if p in token:
                    cutoff = token.index(p)
                    token = token[:cutoff]
            if len(token) > 1:
                entities.append(token)
    if len(entities) > 0:
        return entities
    else:
        return None
entities = [extract_entities(sentence) for sentence in
sentences]
```

```
df = pd.DataFrame({'sentence':sentences, 'entities':entities})
df = df.dropna()
df = df[df['entities'].apply(len) > 1]
entities = df['entities'].to_list()
entities

...
[['Gregor', 'Samsa'],
 ['Oh', 'God'],
 ['God', 'Heaven'],
 ['Gregor', 'Gregor'],
 ['They', 'Gregor'],
 ['Gregor', 'Gregor'],
 ['Yes', 'Gregor'],
 ['Mr', 'Samsa'],
 ['He', 'Gregor'],
 ['Well', 'Mrs', 'Samsa'],
 ['No', 'Gregor'],
 ['Mr', 'Samsa'],
 ['Mr', 'Samsa'],
 ['Sir', 'Gregor'],
 ['Oh', 'God']]
```

This is good enough! We could add a little more logic to prevent the same token from appearing twice in a row, but we can very easily remove those from a network, so let's refactor our code and move on:

```
def get_book_entities():
    text = get_data()
    sentences = sent_tokenize(text)
    entities = [extract_entities(sentence) for sentence in
sentences]
    df = pd.DataFrame({'sentence':sentences,
'entities':entities})
    df = df.dropna()
    df = df[df['entities'].apply(len) > 1]
    entities = df['entities'].to_list()
    return entities
entities = get_book_entities()
entities[0:5]
```

```
...
[['Gregor', 'Samsa'],
 ['Oh', 'God'],
 ['God', 'Heaven'],
 ['Gregor', 'Gregor'],
 ['They', 'Gregor']]
```

This is excellent. At this point, no matter whether we do PoS tagging or NER, we want an entity list, and this is close enough. Next, we will do the same using spaCy, and you should be able to see how spaCy is much easier in some regards. However, there is more setup involved, as you need to install a language model to work with spaCy. There are pros and cons to everything.

Using spaCy for PoS tagging and NER

In this section, I am going to explain how to do what we have just done with NLTK, but this time using spaCy. I will also show you how to use NER as an often-superior alternative to PoS tagging for identifying and extracting entities. Before I started working on this chapter, I primarily used NLTK's PoS tagging as the heart of my entity extraction, but since writing the code for this section and exploring a bit more, I have come to realize that spaCy has improved quite a bit, so I do think that what I am about to show you in this section is superior to what I previously did with NLTK. I do think that it is helpful to explain the usefulness of NLTK. Learn both and use what works best for you. But for entity extraction, I believe spaCy is superior in terms of ease of use and speed of delivery.

Previously, I wrote a function to load Franz Kafka's book *The Metamorphosis*, so we will use that loader here as well, as it has no dependence on either NLTK or spaCy, and it can be easily modified to load any book from the Project Gutenberg archive.

To do anything with spaCy, the first thing you need to do is load the spaCy language model of choice. Before we can load it, we must first install it. You can do so by running the following command:

```
python -m spacy download en_core_web_md
```

There are several different models available, but the three that I use for English text are small, medium, and large. md stands for medium. You could swap that out for sm or lg to get the small or large models, respectively. You can find out more about spaCy's models here: https://spacy.io/usage/models.

Once the model has been installed, we can load it into our Python scripts:

```
import spacy
nlp = spacy.load("en_core_web_md")
```

As mentioned previously, you could swap that md for sm or lg, depending on what model you want to use. The larger model requires more storage and memory for use. The smaller model requires less. Pick the one that works well enough for what you are doing. You may not need the large model. The medium and small ones work well, and the difference between models is often unnoticeable.

Next, we need some text, so let's reuse the function that we previously wrote:

```python
def get_data():
    url = 'https://www.gutenberg.org/files/5200/5200-0.txt'
    text = requests.get(url).text
    # strip header junk
    cutoff = text.index('One morning')
    text = text[cutoff:]
    # strip footer junk
    cutoff = text.rindex('*** END OF THE PROJECT GUTENBERG
EBOOK METAMORPHOSIS ***')
    text = text[:cutoff]
    # pre-processing to clean the text
    text = text.replace('\r', ' ').replace('\n', ' ')
    text = text.replace('â\x80\x99', '\'').replace('â\x80\x9c',
'"').replace('â\x80\x9d', '"').replace('â\x80\x94', ' ')
    return text
```

```
That looks good. We are loading The Metamorphosis, cleaning out
header and footer junk, and then returning text that is clean
enough for our purposes. Just to lay eyes on the text, let's
call our function and inspect the returned text.
text = get_data()
text[0:279]
```

```
...
```

```
'One morning, when Gregor Samsa woke from troubled dreams, he
found himself transformed in his bed into a horrible vermin.
He lay on his  armour-like back, and if he lifted his head
a little he could see his  brown belly, slightly domed and
divided by arches into stiff sections.'
```

There are a few extra spaces in the middle of sentences, but that won't present any problem for us at all. Tokenization will clean that up without any additional effort on our part. We will get to that shortly. First, as done with NLTK, we need to split the text into sentences so that we can build a network based on the entities that are uncovered in each sentence. That is much easier to do using spaCy rather than NLTK, and here is how to do it:

```
doc = nlp(text)
sentences = list(doc.sents)
```

The first line feeds the full text of *The Metamorphosis* into spaCy and uses our language model of choice, while the second line extracts the sentences that are in the text. Now, we should have a Python list of sentences. Let's inspect the first six sentences in our list:

```
for s in sentences[0:6]:
    print(s)
    print()
...
One morning, when Gregor Samsa woke from troubled dreams, he
found  himself transformed in his bed into a horrible vermin.

He lay on his  armour-like back, and if he lifted his head
a little he could see his  brown belly, slightly domed and
divided by arches into stiff sections.

The bedding was hardly able to cover it and seemed ready to
slide off  any moment.

His many legs, pitifully thin compared with the size of
the  rest of him, waved about helplessly as he looked.

"What's happened to me?"

"  he thought.
```

Six probably seems like a strange number of sentences to inspect, but I wanted to show you something. Take a look at the last two sentences. SpaCy has successfully extracted the main character's inner dialog as a standalone sentence, as well as created a separate sentence to complete the surrounding sentence. For our entity extraction, it wouldn't present any problem at all if those sentences were combined, but I like this. It's not a bug, it's a feature, as we software engineers like to say.

SpaCy PoS tagging

Now that we have our sentences, let's use spaCy's PoS tagging as pre-processing for entity extraction:

```
for token in sentences[0]:
    print('{}: {}'.format(token.text, token.tag_))
...
One: CD
morning: NN
,: ,
when: WRB
Gregor: NNP
Samsa: NNP
woke: VBD
from: IN
troubled: JJ
dreams: NNS
,: ,
he: PRP
found: VBD
 :
himself: PRP
transformed: VBD
in: IN
his: PRP$
bed: NN
into: IN
a: DT
horrible: JJ
vermin: NN
.: .
```

Nice. What we want are the NNPs as these are proper nouns. You can see this if you use pos_ instead of tag_:

```
for token in sentences[0]:
    print('{}: {}'.format(token.text, token.pos_))
...
```

```
One: NUM
morning: NOUN
,: PUNCT
when: ADV
Gregor: PROPN
Samsa: PROPN
woke: VERB
from: ADP
troubled: ADJ
dreams: NOUN
,: PUNCT
he: PRON
found: VERB
 : SPACE
himself: PRON
transformed: VERB
in: ADP
his: ADJ
bed: NOUN
into: ADP
a: DET
horrible: ADJ
vermin: NOUN
.: PUNCT
```

Let's add a little logic for extracting them. We need to do two things – we need a list to store the results, and we need some logic to extract the NNPs:

```
entities = []
for token in sentences[0]:
    if token.tag_ == 'NNP':
        entities.append(token.text)
entities

...

['Gregor', 'Samsa']
Perfect! But this is only working on a single sentence. Let's
do the same for all sentences.
```

```
entities = []
for sentence in sentences:
    sentence_entities = []
    for token in sentence:
        if token.tag_ == 'NNP':
            sentence_entities.append(token.text)
    entities.append(sentence_entities)
entities[0:10]
...
[['Gregor', 'Samsa'], [], [], [], [], [], [], [], [],
['Samsa']]
```

For NLTK, we created a function that would extract entities for a given sentence, but this way, I just did everything in one go, and it was quite simple. Let's convert this into a function so that we can easily use this for other future work. Let's also prevent empty lists from being returned inside the entity list, as we have no use for those:

```
def extract_entities(text):
    doc = nlp(text)
    sentences = list(doc.sents)
    entities = []
    for sentence in sentences:
        sentence_entities = []
        for token in sentence:
            if token.tag_ == 'NNP':
                sentence_entities.append(token.text)
        if len(sentence_entities) > 0:
            entities.append(sentence_entities)
    return entities
```

We should now have a clean entity list, with no empty inner lists included:

```
extract_entities(text)
...
[['Gregor', 'Samsa'],
 ['Samsa'],
 ['Gregor'],
 ['God'],
 ['Travelling'],
```

```
['God', 'Heaven'],
['Gregor'],
['Gregor'],
['Gregor'],
['Gregor'],
['Gregor']…]
```

This looks a lot better than NLTK's results, and with fewer steps. This is simple and elegant. I didn't need to use `Pandas` for anything, drop empty rows, or clean out any punctuation that somehow slipped in. We can use this function on any text we have, after cleaning. You can use it before cleaning, but you'll end up getting a bunch of junk entities, especially if you use it against scraped web data.

SpaCy NER

SpaCy's NER is equally easy and simple. The difference between `PoS tagging` and NER is that NER goes a step further and identifies people, places, things, and more. For a detailed description of spaCy's linguistic features, I heartily recommend Duygu Altinok's book *Mastering spaCy*. To be concise, spaCy labels tokens as one of *18* different kinds of entities. In my opinion, that is a bit excessive, as MONEY is not an entity, but I just take what I want. Please check spaCy for the full list of entity types. What we want are entities that are labeled as PERSON, ORG, or GPE. ORG stands for organization, and GPE contains countries, cities, and states.

Let's loop through all the tokens in the first sentence and see how this looks in practice:

```
for token in sentences[0]:
    print('{}: {}'.format(token.text, token.ent_type_))
…
One: TIME
morning: TIME
, :
when:
Gregor: PERSON
Samsa: PERSON
woke:
from:
troubled:
dreams:
, :
he:
found:
```

```
    :
himself:
transformed:
in:
his:
bed:
into:
a:
horrible:
vermin:
 .:
```

This works, but there is a slight problem: we want Gregor Samsa to appear as one entity, not as two. What we need to do is create a new spaCy doc and then loop through the doc's ents rather than the individual tokens. In that regard, the NER approach is slightly different from PoS tagging:

```
doc = nlp(sentences[0].text)


for ent in doc.ents:
    print('{}: {}'.format(ent, ent.label_))
...
One morning: TIME
Gregor Samsa: PERSON
```

Perfect! Let's do a few things: we will redo our previous entity extraction function, but this time using NER rather than PoS tagging, and then limit our entities to PERSON, ORG, and GPE. Please note that I am only adding entities if there is more than one in a sentence. We are looking to identify relationships between people, and you need at least two people to have a relationship:

```
def extract_entities(text):
    doc = nlp(text)
    sentences = list(doc.sents)
    entities = []
    for sentence in sentences:
        sentence_entities = []
        sent_doc = nlp(sentence.text)
        for ent in sent_doc.ents:
            if ent.label_ in ['PERSON', 'ORG', 'GPE']:
                entity = ent.text.strip()
```

```
            if "'s" in entity:
                cutoff = entity.index("'s")
                entity = entity[:cutoff]
            if entity != '':
                sentence_entities.append(entity)
        sentence_entities = list(set(sentence_entities))
        if len(sentence_entities) > 1:
            entities.append(sentence_entities)
    return entities
```

I added a bit of code to remove any whitespace that has snuck in and also to remove duplicates from each sentence_entity list. I also removed any 's characters that appeared after a name – for example, Gregor's – so that it'd show up as Gregor. I could have cleaned this up in network cleanup, but this is a nice optimization. Let's see how our results look. I named the entity list morph_entities for metaMORPHosis entities. I wanted a descriptive name, and this is the best I could come up with:

```
morph_entities = extract_entities(text)
morph_entities

...

[['Gregor', 'Grete'],
 ['Gregor', 'Grete'],
 ['Grete', 'Gregor'],
 ['Gregor', 'Grete'],
 ['Grete', 'Gregor'],
 ['Grete', 'Gregor'],
 ['Grete', 'Gregor'],
 ['Samsa', 'Gregor'],
 ['Samsa', 'Gregor'],
 ['Samsa', 'Grete'],
 ['Samsa', 'Grete'],
 ['Samsa', 'Grete']]
```

That looks great! Wow, I haven't read *The Metamorphosis* in many years and forgot how few characters were in the story!

SpaCy's NER is done using a pre-trained deep learning model, and machine learning is never perfect. Please keep that in mind. There is always some cleanup. SpaCy allows you to customize their language models for your documents, which is useful if you work within a specialized domain, but I prefer to use spaCy's models as general-purpose tools, as I deal with a huge variety of different text. I don't want to customize spaCy for tweets, literature, disinformation, and news. I would prefer to just use it as is and clean up as needed. That has worked very well for me.

You should be able to see that there is a lot of duplication. Apparently, in *The Metamorphosis*, a lot of time is spent talking about Gregor. We will be able to remove those duplicates with a single line of NetworkX code later, so I'll just leave them in rather than tweaking the function. Good enough is good enough. If you are working with massive amounts of data and paying for cloud storage, you should probably fix inefficiencies.

For the rest of this chapter, I'm going to use the NER results as our network data. We could just as easily use the pos_tag entities, but this is better, as NER can combine first name and last name. In our current entities, none of those came through, but they will with other text. This is just how *The Metamorphosis* was written. We'll just clean that up as part of the network creation.

Just for a sanity check, let's check the entities from *Alice's Adventures in Wonderland*!

```python
def get_data():
    url = 'https://www.gutenberg.org/files/11/11-0.txt'
    text = requests.get(url).text
    # strip header junk
    cutoff = text.index('Alice was beginning')
    text = text[cutoff:]
    # strip footer junk
    cutoff = text.rindex('THE END')
    text = text[:cutoff]
    # pre-processing to clean the text
    text = text.replace('\r', ' ').replace('\n', ' ')
    text = text.replace('â\x80\x99', '\'').replace('â\x80\x9c',
'"').replace('â\x80\x9d', '"').replace('â\x80\x94', ' ')
    return text
text = get_data()
text[0:310]
```

...

```
'Alice was beginning to get very tired of sitting by her sister
on the  bank, and of having nothing to do: once or twice she
had peeped into  the book her sister was reading, but it had
```

```
no pictures or  conversations in it, "and what is the use of a
book,"" thought Alice  "without pictures or conversations?""  '
```

I agree with you, Alice.

I have tweaked the loading function to load the book *Alice's Adventures in Wonderland* and chop off any header or footer text. That's actually kind of funny. OFF WITH THEIR HEAD(er)S! Let's try extracting entities. I expect this to be a bit messy, as we are working with fantasy characters, but let's see what happens. Something by Jane Austen might give better results. We'll see!

```
alice_entities = extract_entities(text)
alice_entities[0:10]

...

[['Alice', 'Rabbit'],
 ['Alice', 'Longitude'],
 ['New Zealand', "Ma'am", 'Australia'],
 ['Fender', 'Alice'],
 ['Alice', 'Rabbit'],
 ['Mabel', 'Ada'],
 ['Rome', 'Paris', 'London'],
 ['Improve', 'Nile'],
 ['Alice', 'Mabel'],
 ['Alice', 'William the Conqueror']]
```

That is much better than I expected, but some junk has snuck in. We'll use both of these entity lists for network creation and visualization. This is a good foundation for our next steps. Now that we have some pretty useful-looking entities, let's work toward creating a Pandas DataFrame that we can load into a NetworkX graph! That's what's needed to convert an entity list into an actual social network.

That concludes our demonstration on using spaCy for both PoS tagging as well as NER. I hope you can see that although there was one additional dependency (the language model), the process of entity extraction was much simpler. Now, it is time to move on to what I think is the most exciting part: converting entity lists into network data, which is then used to create a social network, which we can visualize and investigate.

Converting entity lists into network data

Now that we have pretty clean entity data, it is time to convert it into a Pandas DataFrame that we can easily load into NetworkX for creating an actual social network graph. There's a bit to unpack in that sentence, but this is our workflow:

1.   Load text.

2.   Extract entities.

3.   Create network data.

4.   Create a graph using network data.

5.   Analyze the graph.

Again, I use the terms graph and network interchangeably. That does cause confusion, but I did not come up with the names. I prefer to say "network," but then people think I am talking about computer networks, so then I have to remind them that I am talking about graphs, and they then think I am talking about bar charts. You just can't win when it comes to explaining graphs and networks to those who are not familiar, and even I get confused when people start talking about networks and graphs. Do you mean nodes and edges, or do you mean TCP/IP and bar charts? Oh well.

For this next part, we do have choices in how we implement this, but I will explain my typical method. Look at the entities from *Alice's Adventures in Wonderland*:

```
alice_entities[0:10]
...
[['Alice', 'Rabbit'],
 ['Alice', 'Longitude'],
 ['New Zealand', "Ma'am", 'Australia'],
 ['Fender', 'Alice'],
 ['Alice', 'Rabbit'],
 ['Mabel', 'Ada'],
 ['Rome', 'Paris', 'London'],
 ['Improve', 'Nile'],
 ['Alice', 'Mabel'],
 ['Alice', 'William the Conqueror']]
```

In most of these, there are only two entities in each inner list, but sometimes, there are three or more. What I usually do is consider the first entity as the source and any additional entities as targets. What does that look like in a sentence? Let's take this sentence: *"Jack and Jill went up the hill to say hi to their friend Mark."* If we converted that into entities, we would have this list:

```
['Jack', 'Jill', 'Mark']
```

To implement my approach, I will take the first element of my list and add it to my sources list, and then take everything after the first element and add that to my target list. Here is what that looks like in code, but using entities from *Alice*:

```
final_sources = []
final_targets = []
for row in alice_entities:
    source = row[0]
    targets = row[1:]
    for target in targets:
        final_sources.append(source)
        final_targets.append(target)
```

Take a close look at the two lines that capture both `source` and `targets`. `source` is the first element of each entity list, and `targets` is everything after the first element of each entity list. Then, for each target, I add the source and target to `final_sources` and `final_targets`. I loop through `targets` because there can be one or more of them. There will never be more than one `source`, as it is the first element. This is important to understand because this procedure is crucial for how relationships are shown in the resulting social network. We could have used an alternative approach of linking each entity to the other, but I prefer my shown approach. Later lists may bridge any gaps if there is evidence of those relationships. How do our final sources look?

```
final_sources[0:5]
...
['Alice', 'Alice', 'New Zealand', 'New Zealand', 'Fender']
Nice. How about our final targets?
final_targets[0:5]
...
['Rabbit', 'Longitude', "Ma'am", 'Australia', 'Alice']
```

Both look great. Remember, we used NER to capture people, places, and things, so this looks fine. Later, I will very easily drop a few of these sources and targets directly from the social network. This is good enough for now.

The approach of taking the first element and linking it to targets is something that I still regularly consider. Another approach would be to take every entity that appears in the same sentence and link them together. I prefer my approach, but you should consider both options.

The first entity interacts with other entities from the same sentence, but it is not always the case that all entities interact with each other. For instance, look at this sentence:

"John went to see his good friend Aaron, and then he went to the park with Jake."

This is what the entity list would look like:

```
['John', 'Aaron', 'Jake']
```

In this example, Aaron may know Jake, but we can't tell for certain based on this sentence. The hope is that if there is a relationship, that will be picked up eventually. Maybe in another sentence, such as this one:

"Aaron and Jake went ice skating and then ate some pizza."

After that sentence, there will be a definite connection. My preferred approach requires further evidence before connecting entities.

We now have code to take an entity list and create two lists: final_sources and final_targets, but this isn't practical for feeding to NetworkX to create a graph. Let's do two more things: use these two lists to create a Pandas DataFrame, and then create a reusable function that takes any entity list and returns this DataFrame:

```python
def get_network_data(entities):
    final_sources = []
    final_targets = []
    for row in entities:
        source = row[0]
        targets = row[1:]
        for target in targets:
            final_sources.append(source)
            final_targets.append(target)
    df = pd.DataFrame({'source':final_sources, 'target':final_
targets})
    return df
```

That looks great. Let's see it in action!

```python
alice_network_df = get_network_data(alice_entities)
alice_network_df.head()
```

This will display a DataFrame of network data consisting of source and target nodes. This is called an edge list:

	source	target
0	Alice	Rabbit
1	Alice	Longitude
2	New Zealand	Ma'am
3	New Zealand	Australia
4	Fender	Alice

Figure 4.5 – Pandas DataFrame of Alice in Wonderland entity relationships

Great. How does it do with our entities from *The Metamorphosis*?

```
morph_network_df = get_network_data(morph_entities)
morph_network_df.head()
```

This will display the network edge list for *The Metamorphosis*:

	source	target
0	Gregor	Grete
1	Gregor	Grete
2	Grete	Gregor
3	Gregor	Grete
4	Grete	Gregor

Figure 4.6 – Pandas DataFrame of The Metamorphosis entity relationships

Perfect, and the function is reusable!

We are now ready to convert both of these into actual NetworkX graphs. This is where things get interesting, in my opinion. Everything we did previously was just pre-processing. Now, we get to play with networks and specifically social network analysis! After this chapter, we will primarily be learning about social network analysis and network science. There are some areas of NLP that I blew past, such as lemmatization and stemming, but I purposefully did so because they are less relevant to extracting entities than PoS tagging and NER. I recommend that you check out Duygu Altinok's book *Mastering spaCy* if you want to go deeper into NLP. This is as far as we will go with NLP in this book because it is all we need.

Converting network data into networks

It is time to take our created network data and create two graphs, one for *Alice's Adventures in Wonderland*, and another for *The Metamorphosis*. We aren't going to dive deep into network analysis yet, as that is for later chapters. But let's see how they look and see what insights emerge.

First, we need to import the NetworkX library, and then we need to create our graphs. This is extremely easy to do because we have created Pandas DataFrames, which NetworkX will use. This is the easiest way I have found of creating graphs.

First, if you haven't done so yet, you need to install NetworkX. You can do so with the following command:

```
pip install networkx
```

Now that we have installed NetworkX, let's create our two networks:

```
import networkx as nx
G_alice = nx.from_pandas_edgelist(alice_network_df)
G_morph = nx.from_pandas_edgelist(morph_network_df)
```

It's that easy. We have already done the difficult work in our text pre-processing. Did it work? Let's peek into each graph:

```
nx.info(G_alice)

…

'Graph with 68 nodes and 71 edges'

…

nx.info(G_morph)

…

'Graph with 3 nodes and 3 edges'
```

Nice! Already, we are gaining insights into the differences between both social networks. In a graph, a node is just a thing that has relationships with other nodes, typically. Nodes without any relationships are called isolates, but due to the way our graphs have been constructed, there will be no isolates. It's not possible, as we looked for sentences with two or more entities. Try to take a mental picture of those two entities as dots with a line between them. That's literally what a graph/network visualization looks like, except that there are typically many dots and many lines. The relationship that exists between two nodes is called an edge. You will need to understand the difference between nodes and edges to work on graphs.

Looking at the summary information about the *Alice* graph, we can see that there are 68 nodes (characters) and 71 edges (relationships between those characters).

Looking at the summary information about the network from *The Metamorphosis*, we can see that there are only three nodes (characters) and three edges (relationships between those characters. When visualized, this is going to be a really basic network to look at, so I am glad that we did *Alice* as well.

There are many other useful metrics and summaries tucked away inside NetworkX, and we will discuss those when we go over centralities, shortest paths, and other social network analysis and network science topics.

Doing a network visualization spot check

Let's visualize these networks, take a brief look, and then complete this chapter.

Here are two visualization functions that I frequently use. In my opinion, **sknetwork** is superior to NetworkX for network visualization. The visualizations look better, and they render faster. The downside is that there is less customizability, so if you want flexibility, you will need to use NetworkX's native visualizations. Since finding `sknetwork`, I have not looked back to NetworkX for visualization.

The first function converts a NetworkX graph into an adjacency matrix, which sknetwork uses to calculate `PageRank` (an importance score) and then to render the network as an SVG image. The second function uses the first function, but the goal is to visualize an `ego_graph`, which will be described later. In an ego graph, you explore the relationships that exist around a single node. The first function is more general-purpose.

Enough talk. This will be more understandable when you see the results:

```
def draw_graph(G, show_names=False, node_size=1, font_size=10,
edge_width=0.5):
    import numpy as np
    from IPython.display import SVG
    from sknetwork.visualization import svg_graph
    from sknetwork.data import Bunch
    from sknetwork.ranking import PageRank
    adjacency = nx.to_scipy_sparse_matrix(G, nodelist=None,
dtype=None, weight='weight', format='csr')
    names = np.array(list(G.nodes()))
    graph = Bunch()
    graph.adjacency = adjacency
    graph.names = np.array(names)
    pagerank = PageRank()
    scores = pagerank.fit_transform(adjacency)
    if show_names:
        image = svg_graph(graph.adjacency, font_size=font_size,
```

```
    node_size=node_size, names=graph.names, width=700, height=500,
    scores=scores, edge_width=edge_width)
        else:
            image = svg_graph(graph.adjacency, node_size=node_size,
    width=700, height=500, scores = scores, edge_width=edge_width)
        return SVG(image)
```

Next, let's create a function for displaying ego graphs.

To be clear, having `import` statements inside a function is not ideal. It is best to keep import statements external to functions. However, in this case, it makes it easier to copy and paste into your various Jupyter or Colab notebooks, so I am making an exception:

```
def draw_ego_graph(G, ego, center=True, k=0, show_names=True,
    edge_width=0.1, node_size=3, font_size=12):
        ego = nx.ego_graph(G, ego, center=center)
        ego = nx.k_core(ego, k)
        return draw_graph(ego, node_size=node_size, font_size=font_
    size, show_names=show_names, edge_width=edge_width)
```

Look closely at these two functions. Pick them apart and try to figure out what they are doing. To quickly complete this chapter, I'm going to show the results of using these functions. I have abstracted away the difficulty of visualizing these networks so that you can do this:

```
draw_graph(G_alice)
```

As we are not passing any parameters to the function, this should display a very simple network visualization. It will look like a bunch of dots (nodes), with some dots connected to other dots by a line (edge):

Important
Please keep the `draw_graph` function handy. We will use it throughout this book.

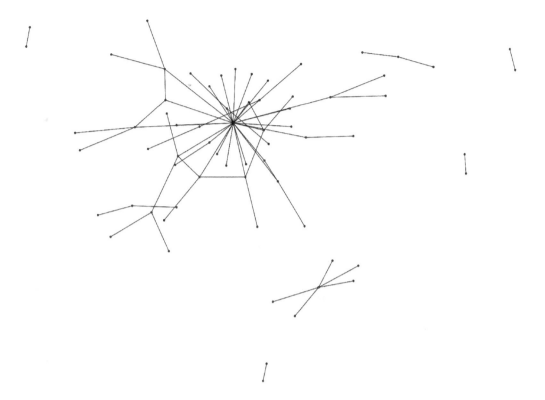

Figure 4.7 – Rough social network of Alice's Adventures in Wonderland

Well, that's a bit unhelpful to look at, but this is intentional. I typically work with large networks, so I prefer to keep node names left out at first so that I can visually inspect the network. However, you can override the default values I am using. Let's do that as well as decrease the line width a bit, increase the node size, and add node names:

```
draw_graph(G_alice, edge_width=0.2, node_size=3, show_
names=True)
```

This will draw our social network, with labels!

Figure 4.8 – Labeled social network of Alice's Adventures in Wonderland

The font size is a bit small and difficult to read, so let's increase that and reduce node_size by one:

```
draw_graph(G_alice, edge_width=0.2, node_size=2, show_
names=True, font_size=12)
```

This creates the following network:

Figure 4.9 – Finalized social network of Alice in Wonderland

That is excellent. Consider what we have done. We have taken raw text from *Alice*, extracted all entities, and built the social network that is described in this book. This is so powerful, and it also opens the door for you to learn more about social network analysis and network science. For instance, would you rather analyze somebody else's toy dataset, or would you rather investigate something you are interested in, such as your favorite book? I prefer to chase my own curiosity.

Let's see what the ego graph looks like around Alice!

```
draw_ego_graph(G_alice, 'Alice')
```

This gives us the following network:

Figure 4.10 – Alice ego graph

What?! That's incredible. We can see that some trash came through in the entity list, but we'll learn how to clean that up in the next chapter. We can take this one step further. What if we want to take Alice out of her ego graph and just explore the relationships that exist around her? Is that possible?

```
draw_ego_graph(G_alice, 'Alice', center=False)
```

This gives us the following visualization:

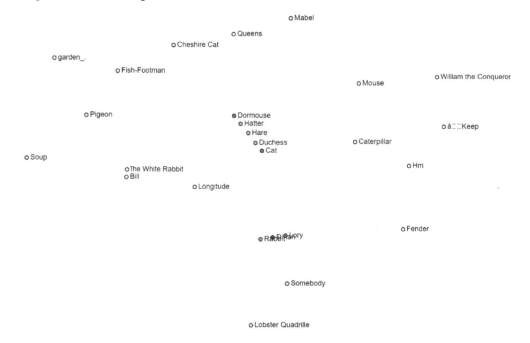

Figure 4.11 – Alice ego graph with dropped center

Too easy. But it's difficult to analyze the clusters of groups that exist. After dropping the center, many nodes became isolates. If only there were a way to remove the isolates so that we could more easily see the groups. OH, WAIT!

```
draw_ego_graph(G_alice, 'Alice', center=False, k=1)
```

We get the following output:

○ Bill   ○ The White Rabbit

● Cat

○ Duchess

○ Hare

○ Hatter

● Dormouse

● Lory

● Dinah

○ Rabbit

Figure 4.12 – Alice ego graph with dropped center and dropped isolates

Overall, the *Alice* social network looks pretty good. There's some cleanup to do, but we can investigate relationships. What does the social network of *The Metamorphosis* look like? Remember, there are only three nodes and three edges. Even Alice's ego graph is more complicated than the social network from *The Metamorphosis*. Let's visualize it!

```
draw_graph(G_morph, show_names=True, node_size=3, font_size=12)
```

This code produces the following network:

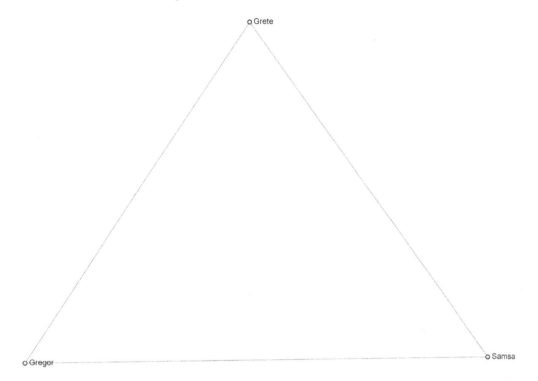

Figure 4.13 – Labeled social network of The Metamorphosis

Wait, but why are there six edges? I only see three. The reason is that sknetwork will draw multiple edges as a single edge. We do have options, such as increasing the line width according to the number of edges but let's just look at the Pandas DataFrame to make sure my thinking is correct:

```
morph_network_df
```

This gets us the following DataFrame:

	source	target
0	Grete	Gregor
1	Grete	Gregor
2	Grete	Gregor
3	Grete	Gregor
4	Grete	Gregor
5	Grete	Gregor
6	Grete	Gregor
7	Samsa	Gregor
8	Samsa	Gregor
9	Grete	Samsa
10	Grete	Samsa
11	Grete	Samsa

Figure 4.14 – Pandas DataFrame of network data for The Metamorphosis

What happens if we drop duplicates?

```
morph_network_df.drop_duplicates()
```

We get the following DataFrame:

	source	target
0	Grete	Gregor
7	Samsa	Gregor
9	Grete	Samsa

Figure 4.15 – Pandas DataFrame of network data for The Metamorphosis (dropped duplicates)

Aha! There is a relationship between Gregor and Grete, but the reverse is also true. One thing that I can see is that Samsa links to Gregor and Grete, but Grete does not link back to Samsa. Another way of saying this, which we will discuss in this book, is that directionality also matters. You can have a directed graph. In this case, I am just using an undirected graph, because relationships are often (but not always) reciprocal.

This marks the end of this demonstration. We originally set out to take raw text and use it to create a social network, and we easily accomplished our goals. Now, we have network data to play with. Now, this book is going to get more interesting.

Additional NLP and network considerations

This has been a marathon of a chapter. Please bear with me a little longer. I have a few final thoughts that I'd like to express, and then we can conclude this chapter.

Data cleanup

First, if you work with language data, there will always be cleanup. Language is messy and difficult. If you are only comfortable working with pre-cleaned tabular data, this is going to feel very messy. I love that, as every project allows me to improve my techniques and tactics.

I showed two different approaches for extracting entities: PoS tagging and NER. Both approaches work very well, but consider which approach gets us closer to a clean and useful entity list the quickest and easiest. With `PoS tagging`, we get one token at a time. With NER, we very quickly get to entities, but the models occasionally misbehave or don't catch everything, so there is always cleanup with this as well.

There is no silver bullet. I want to use whatever approach gets me as close to the goal as quickly as possible because cleanup is inevitable. The less correction I have to do, the quicker I am playing with and pulling insights out of networks.

Comparing PoS tagging and NER

PoS tagging can involve extra steps, but cleanup is often easier. On the other hand, NER can involve fewer steps, but you can get mangled results if you use it against scraped web text. There may be fewer steps for some things, but the cleanup may be daunting. I have seen spaCy's NER false a lot on scraped web content. If you are dealing with web text, spend extra time on cleanup before feeding the data to NER.

Finally, slightly messy results are infinitely better than no results. This stuff is so useful for enriching datasets and extracting the "who, what, and where" parts of any piece of text.

Scraping considerations

There are also a few things to keep in mind when planning any scraping project. First, privacy. If you are scraping social media text, and you are extracting entities from someone else's text, you are running surveillance on them. Ponder how you would feel if someone did the same to you. Further, if you store this data, you are storing personal data, and there may be legal considerations as well. To save yourself headaches, unless you work in government or law enforcement, it might be wise to just use these techniques against literature and news, until you have a plan for other types of content.

There are also ethical considerations. If you decide to use these techniques to build a surveillance engine, you should consider whether building this is an ethical thing to do. Consider whether it is ethical to run surveillance on random strangers.

Finally, scraping is like browsing a website, automatically, but scrapers can do damage. If you hit a website with a scraper a thousand times in a second, you could accidentally hit it with a **DoS** attack. Get what you need at the pace that you need it. If you are looping through all of the links on a website and then scraping them, add a 1-second delay before each scrape rather than hitting it a thousand times every second. You will be liable if you take down a web server, even by accident.

That was a lot of words just to say that unless you are using this for news or literature, be mindful of what you are doing. For news and literature, this can be revealing and may allow new technologies to be created. For other types of content, think about what you are doing before you jump into the work.

Summary

In this chapter, we learned how to find and scrape raw text, convert it into an entity list, and then convert that entity list into an actual social network so that we can investigate revealed entities and relationships. Did we capture the *who*, *what*, and *where* of a piece of text? Absolutely. I hope you can now understand the usefulness of NLP and social network analysis when used together.

In this chapter, I showed several ways to get data. If you are not familiar with web scraping, then this might seem a bit overwhelming, but it's not so bad once you get started. However, in the next chapter, I will show several easier ways to get data.

5
Even Easier Scraping!

In the previous chapter, we covered the basics of web scraping, which is the act of harvesting data from the web for your uses and projects. In this chapter, we will explore even easier approaches to web scraping and will also introduce you to social media scraping. The previous chapter was very long, as we had a lot to cover, from defining scraping to explaining how the **Natural Language Toolkit (NLTK)**, the `Requests` library, and `BeautifulSoup` can be used to collect web data. I will show simpler approaches to getting useful text data with less cleaning involved. Keep in mind that these easier ways do not necessarily replace what was explained in the previous chapter. When working with data or in software projects and things do not immediately work out, it is useful to have options. But for now, we're going to push forward with a simpler approach to scraping web content, as well as giving an introduction to scraping social media text.

First, we will cover the `Newspaper3k` Python library, as well as the `Twitter V2` Python Library.

When I say that `Newspaper3k` is an easier approach to collecting web text data, that is because the authors of `Newspaper3k` did an amazing job of simplifying the process of collecting and enriching web data. They have done much of the work that you would normally need to do for yourself. For instance, if you wanted to gather metadata about a website, such as what language it uses, what keywords are in a story, or even the summary of a news story, `Newspaper3k` gives that to you. This would otherwise be a lot of work. It's easier because, this way, you don't have to do it from scratch.

Second, you will learn how to use the `Twitter V2` Python Library, because this is a very easy way to harvest tweets from Twitter, and this will be useful for NLP as well as network analysis.

We will be covering the following topics in this chapter:

- Why cover Requests and BeautifulSoup?
- Getting started with Newspaper3k
- Introducing the Twitter Python Library

Technical requirements

In this chapter, we will be using the NetworkX and pandas Python libraries. Both of these libraries should be installed by now, so they should be ready for your use. If they are not installed, you can install Python libraries with the following:

```
pip install <library name>
```

For instance, to install NetworkX, you would do this:

```
pip install networkx
```

We will also be discussing a few other libraries:

- Requests
- BeautifulSoup
- Newspaper3k

Requests should already be included with Python and should not need to be installed.

BeautifulSoup can be installed with the following:

```
pip install beautifulsoup4
```

Newspaper3k can be installed with this:

```
pip install newspaper3k
```

In *Chapter 4*, we also introduced a draw_graph() function that uses both NetworkX and scikit-network. You will need that code whenever we do network visualization. Keep it handy!

You can find all the code for this chapter in this book's GitHub repository: https://github.com/PacktPublishing/Network-Science-with-Python.

Why cover Requests and BeautifulSoup?

We all like it when things are easy, but life is challenging, and things don't always work out the way we want. In scraping, that's laughably common. Initially, you can count on more things going wrong than going right, but if you are persistent and know your options, you will eventually get the data that you want.

In the previous chapter, we covered the Requests Python library, because this gives you the ability to access and use any publicly available web data. You get a lot of freedom when working with Requests. This gives you the data, but making the data useful is difficult and time-consuming. We then used BeautifulSoup, because it is a rich library for dealing with HTML. With BeautifulSoup, you can be more specific about the kinds of data you extract and use from a web resource. For instance, we can easily harvest all of the external links from a website, or even the full text of a website, excluding all HTML.

However, BeautifulSoup doesn't give perfectly clean data by default, especially if you are scraping news stories from hundreds of websites, all of which have different headers and footers. The methods we explored in the previous chapter for using offsets to chop off headers and footers are useful when you are collecting text from only a few websites, and you can easily set up a few rules for dealing with each unique website, but then you have a scalability problem. The more websites you decide to scrape, the more randomness you need to be able to deal with. The headers are different, navigation is likely different, and the languages may be different. The difficulties faced during scraping are part of what makes it so interesting to me.

Introducing Newspaper3k

If you want to do NLP or transform text data into network data for use in Social Network Analysis and Network Science, you need clean data. Newspaper3k gives you the next step after BeautifulSoup, abstracting away even more of the cleanup, giving you clean text and useful metadata with less work. Much of the performance and data cleanup have been abstracted away. Also, since you now understand some approaches to cleaning data from the previous chapter, when you see the cleanliness of Newspaper3k's data, you will probably have a better understanding of what is happening under the hood, and hopefully, you will also be quite impressed and thankful for the work they have done and the time they are saving for you.

For now, we can consider Newspaper3k the "easy way" of collecting text off of the web, but it's easy because it builds off of previous foundations.

You will still likely need to use Requests and BeautifulSoup in your web scraping and content analysis projects. We needed to cover them first. When pursuing a web scraping project, rather than start with the most basic and reinventing every tool we need along the way, perhaps we should figure out where to start by asking ourselves a few questions:

- Can I do this with Newspaper3k?
- No? OK, can I do this with Requests and BeautifulSoup?
- No? OK, can I at least get to some data using Requests?

Wherever you get useful results is where you should start. If you can get everything you need from Newpaper3k, start there. If you can't get what you need with Newspaper3k, but you can with BeautifulSoup, start there. If neither of these approaches works, then you're going to need to use Requests to pull data, and then you'll need to write text cleanup code.

I usually recommend that people start at the basics and only add complexity as needed, but I do not recommend this approach for data collection or when it comes to text scraping. There is little use in reinventing HTML parsers. There is no glory in unnecessary headaches. Use whatever gets you useful data the quickest, so long as it meets your project needs.

What is Newspaper3k?

Newspaper3k is a Python library that is useful for loading web text from news websites. However, unlike with BeautifulSoup, the objective is less about flexibility as it is about getting useful data *quickly*. I am most impressed with Newspaper3k's ability to clean data, as the results are quite pure. I have compared my work using BeautifulSoup and Newspaper3k, and I am very impressed with the latter.

Newspaper3k is not a replacement for BeautifulSoup. It can do some things that BeautifulSoup can do, but not everything, and it wasn't written with flexibility for dealing with HTML in mind, which is where BeautifulSoup is strong. You give it a website, and it gives you the text on that website. With BeautifulSoup, you have more flexibility, in that you can choose to look for only links, paragraphs, or headers. Newspaper3k gives you text, summaries, and keywords. BeautifulSoup is a step below that in abstraction. It is important to understand where different libraries sit in terms of tech stacks. BeautifulSoup is a high-level abstraction library, but it is lower level than Newspaper3k. Similarly, BeautifulSoup is a higher-level library than Requests. It's like the movie *Inception* – there are layers and layers and layers.

What are Newspaper3k's uses?

Newspaper3k is useful for getting to the clean text that exists in a web news story. That means parsing the HTML, chopping out non-useful text, and returning the news story, headline, keywords, and even text summary. The fact that it can return keywords and text summaries means that it's got some pretty interesting NLP capabilities in the background. You don't need to create a machine learning model for text summarization. Newspaper3k will do the work for you transparently, and surprisingly fast.

Newspaper3k seems to have been inspired by the idea of parsing online news, but it is not limited to that. I have also used it for scraping blogs. If you have a website that you'd like to try scraping, give Newspaper3k a chance and see how it does. If that doesn't work, use BeautifulSoup.

One weakness of Newspaper3k is that it is unable to parse websites that use JavaScript obfuscation to hide their content inside JavaScript rather than HTML. Web developers occasionally do this to discourage scraping, for various reasons. If you point Newspaper3k or BeautifulSoup at a website that is using JavaScript obfuscation, both of them will return very little or no useful results, as the data is hidden in JavaScript, which neither of these libraries is built to handle. The workaround is to use a library such as Selenium along with Requests, and that will often be enough to get to the data that you want. Selenium is outside the scope of this book, and often feels like more trouble than it is worth, so please explore the documentation if you get stuck behind JavaScript obfuscation, or just move on and scrape easier websites. Most websites are scrapable, and the ones that aren't can often just be ignored as they may not be worth the effort.

Getting started with Newspaper3k

Before you can use Newspaper3k, you must install it. This is as simple as running the following command:

```
pip install newspaper3k
```

At one point in a previous installation, I received an error stating that an NLTK component was not downloaded. Keep an eye out for weird errors. The fix was as simple as running a command for an NLTK download. Other than that, the library has worked very well for me. Once the installation is complete, you will be able to import it into your Python code and make use of it immediately.

In the previous chapter, I showed flexible but more manual approaches to scraping websites. A lot of junk text snuck through, and cleaning the data was quite involved and difficult to standardize. Newspaper3k takes scraping to another level, making it easier than I have ever seen anywhere else. I recommend that you use Newspaper3k for your news scraping whenever you can.

Scraping all news URLs from a website

Harvesting URLs from a domain using Newspaper3k is simple. This is all of the code required to load all the hyperlinks from a web domain:

```
import newspaper
domain = 'https://www.goodnewsnetwork.org'
paper = newspaper.build(domain, memoize_articles=False)
urls = paper.article_urls()
```

However, there is one thing I want to point out: when you scrape all URLs this way, you will also find what I consider "junk URLs" that point to other areas of the website, not to articles. These can be useful, but in most cases, I just want the article URLs. This is what the URLs will look like if I don't do anything to remove the junk:

```
urls
['https://www.goodnewsnetwork.org/2nd-annual-night-of-a-
million-lights/',
 'https://www.goodnewsnetwork.org/cardboard-pods-for-animals-
displaced-by-wildfires/',
 'https://www.goodnewsnetwork.org/category/news/',
 'https://www.goodnewsnetwork.org/category/news/animals/',
 'https://www.goodnewsnetwork.org/category/news/arts-leisure/',
```

```
'https://www.goodnewsnetwork.org/category/news/at-home/',
'https://www.goodnewsnetwork.org/category/news/business/',
'https://www.goodnewsnetwork.org/category/news/celebrities/',
'https://www.goodnewsnetwork.org/category/news/earth/',
'https://www.goodnewsnetwork.org/category/news/founders-blog/']
```

Please take note that if you crawl a website at a different time, you will likely get different results. New content may have been added, and old content may have been removed.

Everything under those first two URLs is what I consider junk, in most of my scraping. I want the article URLs, and those are URLs to specific category pages. There are several ways that this problem can be addressed:

- You could drop URLs that include the word "category." In this case, that looks perfect.
- You could drop URLs where the length of the URL is greater than a certain threshold.
- You could combine the two options into a single approach.

For this example, I have decided to go with the third option. I will drop all URLs that include the word "category," as well as any URLs that are less than 60 characters in length. You may want to experiment with various cutoff thresholds to see what works for you. The simple cleanup code looks like this:

```
urls = sorted([u for u in urls if 'category' not in u and
len(u)>60])
```

Our URL list now looks much cleaner, containing only article URLs. This is what we need:

```
urls[0:10]
...
['https://www.goodnewsnetwork.org/2nd-annual-night-of-a-
million-lights/',
 'https://www.goodnewsnetwork.org/cardboard-pods-for-animals-
displaced-by-wildfires/',
 'https://www.goodnewsnetwork.org/couple-living-in-darkest-
village-lights-sky-with-huge-christmas-tree/',
 'https://www.goodnewsnetwork.org/coya-therapies-develop-
breakthrough-treatment-for-als-by-regulating-t-cells/',
 'https://www.goodnewsnetwork.org/enorme-en-anidacion-de-
tortugasen-tailandia-y-florida/',
 'https://www.goodnewsnetwork.org/good-talks-sustainable-dish-
podcast-with-shannon-hayes/',
 'https://www.goodnewsnetwork.org/gopatch-drug-free-patches-
```

```
good-gifts/',
 'https://www.goodnewsnetwork.org/horoscope-from-rob-brezsnys-
free-will-astrology-12-10-21/',
 'https://www.goodnewsnetwork.org/how-to-recognize-the-eight-
forms-of-capital-in-our-lives/',
 'https://www.goodnewsnetwork.org/mapa-antiguo-de-la-tierra-te-
deja-ver-su-evolucion/']
```

We now have a clean URL list that we can iterate through, scrape each story, and load the text for our use.

Before moving on, one thing that you should notice is that on this single web domain, the stories that they publish are multilingual. Most of the stories that they publish are in English, but some of them are not. If you were to point Newspaper3k at the domain (rather than at individual story URLs), it would likely be unable to correctly classify the language of the domain. It is best to do language lookups at the story level, not the domain level. I will show how to do this at the story level.

Scraping a news story from a website

We now have a list of story URLs that we want to scrape article text and metadata from. The next step is to use a chosen URL and harvest any data that we want. For this example, I will download and use the first story URL in our URL list:

```
from newspaper import Article
url = urls[0]
article = Article(url)
article.download()
article.parse()
article.nlp()
```

There are a few confusing lines in this code snippet, so I will explain it line by line:

1. First, I load the Article function from the newspaper library, as that is used for downloading article data.

2. Next, I point Article at the first URL from our URL list, which is urls[0]. It has not done anything at this point; it has just been pointed at the source URL.

3. Then, I download and parse the text from the given URL. This is useful for grabbing full text and headlines, but it will not capture article keywords.

4. Finally, I run the nlp component of Article to extract keywords.

With these four steps, I should now have all the data that I want for this article. Let's dive in and see what we have!

- What is the article title?

```
title = article.title
title
```

...

```
'After Raising $2.8M to Make Wishes Come True for Sick
Kids, The 'Night of a Million Lights' Holiday Tour is
Back'
```

- Nice and clean. What about the text?

```
text = article.text
text[0:500]
```

...

```
'The Night of A Million Lights is back—the holiday
spectacular that delights thousands of visitors and
raises millions to give sick children and their weary
families a vacation.\n\n'Give Kids The World Village' has
launched their second annual holiday lights extravaganza,
running until Jan. 2\n\nIlluminating the Central Florida
skyline, the 52-night open house will once again provide
the public with a rare glimpse inside Give Kids The World
Village, an 89-acre, whimsical nonprofit resort that
provide'
```

- What is the article summary?

```
summary = article.summary
summary
```

...

```
'The Night of A Million Lights is back—the holiday
spectacular that delights thousands of visitors and
raises millions to give sick children and their weary
families a vacation.\nWhat began as an inventive pandemic
pivot for Give Kids The World has evolved into Central
Florida's most beloved new holiday tradition.\n"Last
year's event grossed $2.8 million to make wishes come
true for children struggling with illness and their
families," spokesperson Cindy Elliott told GNN.\nThe
```

```
display features 1.25M linear feet of lights, including
3.2 million lights that were donated by Walt Disney
World.\nAll proceeds from Night of a Million Lights will
support Give Kids The World, rated Four Stars by Charity
Navigator 15 years in a row.'
```

- What language was the article written in?

```
language = article.meta_lang
language

...

'en'
```

- What keywords were found in the article?

```
keywords = article.keywords
keywords

...

['million',
 'kids',
 'children',
 'lights',
 'world',
 'true',
 'tour',
 'wishes',
 'sick',
 'raising',
 'night',
 'village',
 'guests',
 'holiday',
 'wish']
```

- What image accompanies this story?

```
image = article.meta_img
image

...

'https://www.goodnewsnetwork.org/wp-content/
```

```
uploads/2021/12/Christmas-disply-Night-of-a-Million-
Lights-released.jpg'
```

And there is even more that you can do with `Newspaper3k`. I encourage you to read the library's documentation and see what else can be useful to your work. You can read more at `https://newspaper.readthedocs.io/en/latest/`.

Scraping nicely and blending in

There are two things that I try to do when building any scraper:

- Blend in with the crowd

- Don't scrape too aggressively

There is some overlap between both of these. If I blend in with actual website visitors, my scrapers will be less noticeable, and less likely to get blocked. Second, if I don't scrape too aggressively, my scrapers are not likely to be noticed, and thus also less likely to be blocked. However, the second one is important, as it is not friendly to hit web servers too aggressively. It is better to throw in a 0.5 or 1-second wait between URL scrapes:

1. For the first idea, blending in with the crowd, you can spoof a browser user-agent. For instance, if you want your scraper to pretend to be the latest Mozilla browser running on macOS, this is how to do so:

    ```
    from newspaper import Config
    config = Config()
    config.browser_user_agent = 'Mozilla/5.0 (Macintosh;
    Intel Mac OS X 12.0; rv:95.0) Gecko/20100101
    Firefox/95.0'
    config.request_timeout = 3
    ```

2. Next, to add a 1-second sleep between each URL scrape, you could use a `sleep` command:

    ```
    import time
    time.sleep(1)
    ```

The `user_agent` configuration is often enough to get past simple bot detection, and the 1-second sleep is a friendly thing to do that also helps with blending in.

Converting text into network data

To convert our freshly scraped text into network data, we can reuse the function that was created in the previous chapter. As a reminder, this function was created to use an NLP technique called **Named-Entity Recognition** (**NER**) to extract the people, places, and organizations mentioned in a document:

1. Here is the function that we will use:

```python
import spacy
nlp = spacy.load("en_core_web_md")
def extract_entities(text):
    doc = nlp(text)
    sentences = list(doc.sents)
    entities = []
    for sentence in sentences:
        sentence_entities = []
        sent_doc = nlp(sentence.text)

        for ent in sent_doc.ents:
            if ent.label_ in ['PERSON', 'ORG', 'GPE']:
                entity = ent.text.strip()
                if "'s" in entity:
                    cutoff = entity.index("'s")
                    entity = entity[:cutoff]
                if entity != '':
                    sentence_entities.append(entity)
        sentence_entities = list(set(sentence_entities))
        if len(sentence_entities) > 1:
            entities.append(sentence_entities)
    return entities
```

2. We can simply throw our scraped text into this function and it should return an entity list:

```python
entities = extract_entities(text)
entities

...

[['Night', 'USA'],  ['Florida', 'Kissimmee'],  ['GNN',
'Cindy Elliott'],  ['the Centers for Disease Control and
Prevention', 'CDC'],  ['Florida', 'Santa'],  ['Disney
World', 'Central Florida']]
```

3. Perfect! Now, we can pass these entities to one additional function to get a `pandas` DataFrame of edge list data that we can use to construct a network.

4. Next, we will use the `get_network_data` function, which is coded as follows:

```
import pandas as pd
def get_network_data(entities):
    final_sources = []
    final_targets = []
    for row in entities:
        source = row[0]
        targets = row[1:]
        for target in targets:
            final_sources.append(source)
            final_targets.append(target)
    df = pd.DataFrame({'source':final_sources,
'target':final_targets})
    return df
```

5. We can use it by passing in an entity list:

```
network_df = get_network_data(entities)
network_df.head()
```

Upon inspection, this looks great. A network edge list must contain a source node and a target node, and we've now got both in place:

	source	target
0	Night	USA
1	Florida	Kissimmee
2	GNN	Cindy Elliott
3	the Centers for Disease Control and Prevention	CDC
4	Florida	Santa

Figure 5.1 – pandas DataFrame edge list of entities

That's great. The fourth row is interesting, as NER successfully caught two different ways of representing the CDC, both spelled out and as an acronym. There seems to be a false positive in the first row, but I will explain how to clean network data in the next chapter. This is perfect for now.

End-to-end Network3k scraping and network visualization

We now have everything we need to demonstrate two things. I want to show how to scrape several URLs and combine the data into a single DataFrame for use and storage, and you will learn how to convert raw text into network data and visualize it. We did the latter in the previous chapter, but it will be useful to do it one more time so that the learning sticks.

Combining multiple URL scrapes into a DataFrame

Between these two demonstrations, this is the most foundational and important part. We will use the results of this process in our next demonstration. In most real-world scraping projects, it is not useful to scrape a single URL repeatedly. Typically, you want to repeat these steps for any given domain that you scrape:

1. Scrape all URLs.

2. Drop the ones that you have already scraped text for.

3. Scrape the text of the URLs that remain.

For this demonstration, we will only be doing *step 1* and *step 3*. For your projects, you will usually need to come up with a process to drop the URLs you have already scraped, and this is dependent on where you are writing the post-scraped data. Essentially, you need to take a look at what you have and disregard any URLs that you have already used. This prevents repeated work, unnecessary scraping noise, unnecessary scraping burden on web servers, and duplicated data.

The following code scrapes all URLs for a given domain, scrapes text for each URL discovered, and creates a `pandas` DataFrame for use or writing output to a file or a database. I am throwing one additional Python library at this: `tqdm`. The `tqdm` library is useful when you want to understand how long a process will take. If you are using this in backend automation, you will likely not want the `tqdm` functionality, but it is useful now, as you are learning.

You can install `tqdm` by running `pip install tqdm`:

```
domain = 'https://www.goodnewsnetwork.org'

df = get_story_df(domain)
df.head()
```

```
21%|██████████████████          | 5/24 [00:03<00:12,  1.52it/s]
```

Figure 5.2 – TQDM progress bar in action

This is the end-to-end Python code that takes a domain name and returns a pandas DataFrame of scraped stories:

```python
import newspaper
from newspaper import Article
from tqdm import tqdm
def get_story_df(domain):
    paper = newspaper.build(domain, memoize_articles=False)
    urls = paper.article_urls()
    urls = sorted([u for u in urls if 'category' not in u and
len(u)>60])
    titles = []
    texts = []
    languages = []
    keywords = []
    for url in tqdm(urls):
        article = Article(url)
        article.download()
        article.parse()
        article.nlp()
        titles.append(article.title)
        texts.append(article.text)
        languages.append(article.meta_lang)
        keywords.append(article.keywords)
    df = pd.DataFrame({'urls':urls, 'title':titles,
'text':texts, 'lang':languages, 'keywords':keywords})
    return df
```

To use this function, you can run the following code, and point it at any news domain of interest:

```python
domain = 'https://www.goodnewsnetwork.org'
df = get_story_df(domain)
df.head()
```

You should now have a clean DataFrame of news stories to work with. If you are running into 404 (Page Not Found) errors, you may need to place some try/except exception handling code into the function. I leave this and other edge cases in your hands. However, the closer the time is between URL scraping and article text scraping, the less likely you will run into 404 errors.

Let's inspect the results!

```
100%|████████████████████████████████████████████| 24/24 [00:17<00:00,  1.39it/s]
```

	urls	title	text	lang	keywords
0	https://www.goodnewsnetwork.org/2nd-annual-nig...	After Raising $2.8M to Make Wishes Come True f...	The Night of A Million Lights is back—the holi...	en	[million, kids, children, lights, world, true,...
1	https://www.goodnewsnetwork.org/cardboard-pods...	Cardboard Habitat Pods Give a Fighting Chance ...	New habitat pods developed by an Australian Un...	en	[cardboard, habitat, small, fighting, wildfire...
2	https://www.goodnewsnetwork.org/couple-living-...	Elderly Couple Living in UK's Darkest Village ...	This tree-mendous pine is becoming known as Th...	en	[sky, lights, christmas, huge, living, tree, b...
3	https://www.goodnewsnetwork.org/coya-therapies...	New Simple Therapy Offers Potentially Groundbr...	A man with Amyotrophic Lateral Sclerosis (ALS)...	en	[potentially, life, groundbreaking, offers, au...
4	https://www.goodnewsnetwork.org/enorme-en-anid...	Las Playas han Visto un Incremento Enorme en A...	Desde peces volviendo a los canales de Venecia...	es	[en, los, las, playas, la, laúd, restricciones...

Figure 5.3 – pandas DataFrame of scraped URL data

Cool! The `tqdm` progress bar worked until completion, and we can also see that the final story's language was set to Spanish. This is exactly what we want. If we had tried to detect the language of the overall domain, by scraping the landing page (home page), the language detection component might have given a false reading or even returned nothing. This website has both English and Spanish language stories, and we can see this at the story level.

Capturing the language of a piece of text is very useful for NLP work. Often, machine learning classifiers that are trained in one language will have difficulty when used against another, and when using unsupervised machine learning (clustering) against text data, data written in various languages will clump together. My advice is to use the captured language data to split your data by language for any downstream NLP enrichment work. You will have better results this way, and your results will be much simpler to analyze as well.

Next, let's use these stories to create network data and visualizations!

Converting text data into a network for visualization

Several times in this book, we have taken text, extracted entities, created network data, created a network, and then visualized the network. We will be doing the same thing here. The difference is that we now have a `pandas` DataFrame that consists of several news articles, and each of them can be converted into a network. For this example, I'll only do it twice. From this point on, you should have no trouble converting text into networks, and you can reuse the code that has already been written.

Our entity extraction is built upon an English language NLP model, so let's only use English language stories. To keep things simple, we will do this demonstration with the second and fourth articles in the DataFrame, as these gave interesting and clean results:

1. First, we will use the second article. You should see that I am loading `df['text'][1]`, where `[1]` is the second row as indexing starts at 0:

```
text = df['text'][1]
entities = extract_entities(text)
```

```
network_df = get_network_data(entities)
G = nx.from_pandas_edgelist(network_df)
draw_graph(G, show_names=True, node_size=4, edge_width=1,
font_size=12)
```

This is the network visualization:

○ NSW National Parks and Wildlife Service

○ the Australian Wildlife Conservancy

Figure 5.4 – Network visualization of article entity relationships (second article)

This looks good but is very simple. A news article is typically about a few individuals and organizations, so this is not surprising. We can still see a relationship between a couple of wildlife groups, a relationship between a person and a university, and a relationship between **Australia** and **Koalas**. All of this seems realistic.

2. Next, let's try the fourth article:

```
text = df['text'][3]
entities = extract_entities(text)
network_df = get_network_data(entities)
G = nx.from_pandas_edgelist(network_df)
draw_graph(G, show_names=True, node_size=4, edge_width=1,
font_size=12)
```

This is the network visualization. This one is much more interesting and involved:

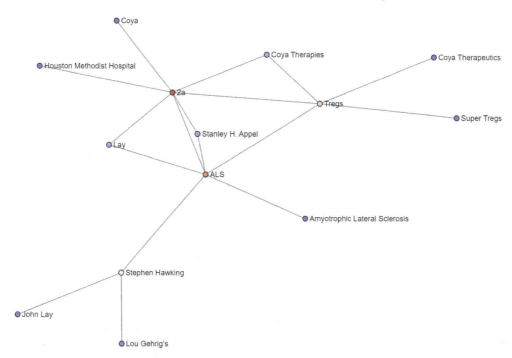

Figure 5.5 – Network visualization of article entity relationships (fourth article)

This is a much richer set of entities than is commonly found in news stories, and in fact, this story has more entities and relationships than the book that we investigated in the previous chapter, *The Metamorphosis*. This looks great, and we can investigate the relationships that have been uncovered.

From this point on in this book, I will primarily be using Twitter data to create networks. I wanted to explain how to do this to any text, as this gives freedom to uncover relationships in any text, not just social media text. However, you should understand this by now. The remainder of this book will focus more on analyzing networks than on creating network data. Once text data has been converted into a network, the rest of the network analysis information is equally relevant and useful.

Introducing the Twitter Python Library

Twitter is a goldmine for NLP projects. It is a very active social network with not too strict moderation, which means that users are pretty comfortable posting about a wide variety of topics. This means that Twitter can be useful for studying lighthearted topics, but it can also be used to study more serious topics. You have a lot of flexibility.

Twitter also has a simple API to work with, compared to other social networks. It is relatively simple to get started with, and it can be used to capture data that can be used for a lifetime of NLP research. In my personal NLP research, learning to scrape Twitter supercharged and accelerated my NLP learning. Learning NLP is much more enjoyable when you have data that is interesting to you. I have used Twitter data to understand various networks, create original NLP techniques, and create machine learning training data. I don't use Twitter much, but I have found it to be a goldmine for all things NLP.

What is the Twitter Python Library?

For several years, Twitter has been exposing its API, to allow software developers and researchers to make use of their data. The API is a bit of a challenge to use as the documentation is a bit scattered and confusing, so a Python library was created to make working with the API much easier. I will explain how to use the Python library, but you will need to explore the Twitter API itself to learn more about the various limits that Twitter has set in place to prevent overuse of the API.

You can read more about the Twitter API at `https://developer.twitter.com/en/docs`.

What are the Twitter Library's uses?

Once you can use the Twitter library and API, you have total flexibility in what you use it for to research. You could use it to learn about the K-Pop music scene, or you could use it to keep an eye on the latest happenings in machine learning or data science.

Another use of this is analyzing entire audiences. For instance, if an account has 50,000 followers, you can use the Twitter Library to load data about all 50,000 of the followers, including their usernames and descriptions. With these descriptions, you could use clustering techniques to identify the various subgroups that exist in a larger group. You could also use this kind of data to potentially identify bots and other forms of artificial amplification.

I recommend that you find something that you are curious about, and then chase it and see where it leads. This curiosity is an excellent driver for building skills in NLP and Social Network Analysis.

What data can be harvested from Twitter?

Even compared to 2 years ago, it seems that Twitter has expanded its API offerings to allow for many different kinds of data science and NLP projects. However, in their **version one** (**V1**) API, Twitter would return a dictionary containing a great deal of the data that they have available. This has changed a bit in their V2 API, as they now require developers to specify which data they are requesting. This has made it more difficult to know all of the data that Twitter has made available. Anyone who will be working with the Twitter API is going to need to spend time reading through the documentation to see what is available.

For my research, I am typically only interested in a few things:

- Who is posting something?

- What have they posted?

- When was it posted?

- Who are they mentioning?

- What hashtags are they using?

All of this is easy to pull from the Twitter API, and I will show you how. But this is not the limit of Twitter's offerings. I have recently been impressed by some of the more recently discovered data that the V2 API exposes, but I do not understand it well enough to write about it, yet. When you run into something that feels like it should be exposed by the API, check the documentation. In my experience working with the Twitter API, some things that should be exposed by default now take a bit of extra work than in V1. Try to figure out how to get what you need.

Getting Twitter API access

Before you can do anything with the Twitter API, the first thing you need to do is get access:

1. First, create a Twitter account. You don't need to use it to post anything, but you do need to have an account.

2. Next, go to the following URL to request API access: `https://developer.twitter.com/en/apply-for-access`.

    Applying for access can vary in time from a few minutes to a few days. You will need to fill out a few forms specifying how you will be using the data and agreeing to abide by Twitter's Terms of Service. In describing your use of Twitter data, you can specify that you are using this for learning NLP and Social Network Analysis.

3. Once you have been granted access, you will have your own Developer Portal. Search around in the authentication section until you see something like this:

Figure 5.6 – Twitter Authentication Bearer Token

Specifically, you are looking for a **Bearer Token**. Generate one and keep it somewhere safe. You will use this to authenticate with the Twitter API.

Once you have generated a Bearer Token, you should be all set to work with the Twitter API through the Twitter Python library.

Authenticating with Twitter

Before you can authenticate, you need to install the Twitter Python library:

1. You can do so by running the following:

    ```
    pip install python-twitter-v2
    ```

2. Next, in your notebook of choice, try importing the library:

    ```
    from pytwitter import Api
    ```

3. Next, you will need to authenticate with Twitter using your Bearer Token. Replace the `bearer_token` text in the following code with your own Bearer Token and try authenticating:

    ```
    bearer_token = 'your_bearer_token'
    twitter_api = Api(bearer_token=bearer_token)
    ```

If this doesn't fail, then you should be authenticated and ready to start scraping tweets, connections, and more.

Scraping user tweets

I've created two helper functions for loading user tweets into a `pandas` DataFrame. If you want more data than this function returns, you will need to extend `tweet_fields`, and possibly add `user_fields`:

1. Please see the `search_tweets()` function in the following code block to see how `user_fields` can be added to a Twitter call. This function does not use `user_fields`, as we are already passing in a username:

    ```
    def get_user_id(twitter_api, username):

        user_data = twitter_api.get_users(usernames=username)

        return user_data.data[0].id
    ```

2. This first function takes a Twitter username and returns its `user_id`. This is important because some Twitter calls require a `user_id`, not a `username`. The following function uses `user_id` to look up a user's tweets:

```
def get_timeline(twitter_api, username):

    tweet_fields = ['created_at', 'text', 'lang']
    user_id = get_user_id(twitter_api, username)

    timeline_data = twitter_api.get_timelines(user_id,
    return_json=True, max_results=100, tweet_fields=tweet_
    fields)

    df = pd.DataFrame(timeline_data['data'])
    df.drop('id', axis=1, inplace=True)

    return df
```

In this function, the `twitter_api.get_timelines()` function is doing most of the work. I have specified `tweet_fields` that I want, I've passed in a user's `user_id`, I've specified that I want the latest `100` tweets by that person, and I've specified that I want the data returned in JSON format, which is easy to convert into a `pandas` DataFrame. If I call this function, I should get immediate results.

3. Let's see what Santa Claus talks about:

```
df = get_timeline(twitter_api, 'officialsanta')
df.head()
```

We should now see a preview of five of Santa's tweets:

	text	lang	created_at
0	Today is..\n😀⭐😀⭐😀⭐😀⭐😀⭐\n 17 SLEEPS TO CH...	en	2021-12-08T08:09:46.000Z
1	RT @s777rts: Thank you @OfficialSanta https://...	en	2021-12-07T10:33:50.000Z
2	RT @seanydisora: My little boy kitted out read...	en	2021-12-07T10:31:04.000Z
3	Official Santa Letters 😀📜\n⭐⭐⭐⭐⭐\nOrder today ...	en	2021-12-04T20:04:55.000Z
4	Today is..\n😀⭐😀⭐😀⭐😀⭐😀⭐\n 23 SLEEPS TO CH...	en	2021-12-02T20:47:24.000Z

Figure 5.7 – pandas DataFrame of Santa Claus tweets

Perfect. We now have the most recent `100` tweets made by Santa Claus.

4. I have one additional helper function that I would like to give you. This one takes the `text` field and extracts entities and hashtags; we'll be using these to draw social networks:

```
def wrangle_and_enrich(df):


    # give some space for splitting, sometimes things get
smashed together
    df['text'] = df['text'].str.replace('http', ' http')
    df['text'] = df['text'].str.replace('@', ' @')
    df['text'] = df['text'].str.replace('#', ' #')
    # enrich dataframe with user mentions and hashtags
    df['users'] = df['text'].apply(lambda tweet:
[clean_user(token) for token in tweet.split() if token.
startswith('@')])
    df['tags'] = df['text'].apply(lambda tweet: [clean_
hashtag(token) for token in tweet.split() if token.
startswith('#')])


    return df
```

5. We can add this function as an enrichment step:

```
df = get_timeline(twitter_api, 'officialsanta')
df = wrangle_and_enrich(df)
df.head()
```

This gives us additional useful data:

	text	created_at	lang	users	tags
0	Today is..\n😂⭐😊⭐😂⭐😊⭐😂⭐\n 17 SLEEPS TO CH...	2021-12-08T08:09:46.000Z	en	[]	[#countdowntochristmas]
1	RT @s777rts: Thank you @OfficialSanta https...	2021-12-07T10:33:50.000Z	en	[@s777rts, @officialsanta]	[]
2	RT @seanydisora: My little boy kitted out rea...	2021-12-07T10:31:04.000Z	en	[@seanydisora, @officialsanta]	[#santadash, #fundraising, #school, #books]
3	Official Santa Letters 😂📩\n⭐ ⭐ ⭐ ⭐ ⭐\nOrder today ...	2021-12-04T20:04:55.000Z	en	[]	[]
4	Today is..\n😂⭐😊⭐😂⭐😊⭐😂⭐\n 23 SLEEPS TO CH...	2021-12-02T20:47:24.000Z	en	[]	[#countdowntochristmas]

Figure 5.8 – pandas DataFrame of enriched Santa Claus tweets

This is perfect for the rest of the work we will be doing in this chapter, but before we move on, we will also see how to scrape connections.

Scraping user following

We can easily scrape all accounts that an account follows. This can be done with the following function:

```
def get_following(twitter_api, username):
    user_fields = ['username', 'description']
    user_id = get_user_id(twitter_api, username)
    following = twitter_api.get_following(user_id=user_id,
return_json=True, max_results=1000, user_fields=user_fields)
    df = pd.DataFrame(following['data'])
    return df[['name', 'username', 'description']]
```

Here, I have specified max_results=1000. That is the maximum that Twitter will return at a time, but you can load much more than 1,000. You will need to pass in a 'next_token' key to continue harvesting sets of 1000 followers. You can do something similar to load more than 100 tweets by a person. Ideally, you should use recursion in programming to do this, if you have the need. You can use the preceding function to load the first batch, and you should be able to extend it if you need to build in recursion.

You can call the following function:

```
df = get_following(twitter_api, 'officialsanta')
df.head()
```

This will give you results in this format:

	name	username	description
0	🎄🎄🎄🎄 🎄	CFCREDYR	Chelsea fan • I support my local Charlton Athl...
1	Cllr Andrew Morgan	AndrewMorganRCT	Cllr Mountain Ash West & Labour Leader @rctcou...
2	Axy 🎄	utdaxy	|| fan account || • @ManUtd •
3	🎄🎄🎄🎄🎄🎄🎄🎄	Carefree195	Regular quality tweets \nUp the Chels\nmake su...
4	Joël Veltman is (not) actually the goat	Lassinaasappel	Aired

Figure 5.9 – pandas DataFrame of accounts Santa Claus follows on Twitter

For investigating subgroups that exist inside a group, it is useful to include the account description, as people are often descriptive about their interests and political affiliations. To capture the description, you need to include the description in the user_fields list.

Scraping user followers

Scraping followers is nearly identical. Here is the code:

```
def get_followers(twitter_api, username):
    user_fields = ['username', 'description']
    user_id = get_user_id(twitter_api, username)
    followers = twitter_api.get_followers(user_id=user_id,
return_json=True, max_results=1000, user_fields=user_fields)
    df = pd.DataFrame(followers['data'])
    return df[['name', 'username', 'description']]
```

You can call the function:

```
df = get_followers(twitter_api, 'officialsanta')
df.head()
```

This will give you results in the same format as was shown previously. Be sure to include the account description.

Scraping using search terms

Collecting tweets about a search term is also useful. You can use this to explore *who* participates in discussions about a search term, but also to collect the tweets themselves, for reading and processing.

Here is the code that I have written for scraping by search term:

```
def search_tweets(twitter_api, search_string):
    tweet_fields = ['created_at', 'text', 'lang']
    user_fields = ['username']
    expansions = ['author_id']
    search_data = twitter_api.search_tweets(search_string,
return_json=True, expansions=expansions, tweet_fields=tweet_
fields, user_fields=user_fields, max_results=100)
    df = pd.DataFrame(search_data['data'])
    user_df = pd.DataFrame(search_data['includes']['users'])
    df = df.merge(user_df, left_on='author_id', right_on='id')
    df['username'] = df['username'].str.lower()
    return df[['username', 'text', 'created_at', 'lang']]
```

This function is a bit more involved than the previous functions, as I have specified `tweet_fields` as well as `user_fields` that I am interested in. To capture the username, I needed to specify an expansion on `author_id`, and finally, I want 100 of the latest tweets. If you want to include additional data, you will need to explore the Twitter API to find out how to add the data field of interest.

You can call the function like so:

```
df = search_tweets(twitter_api, 'natural language processing')
df = wrangle_and_enrich(df)
df.head()
```

I am also enriching the `pandas` DataFrame so that it includes user mentions and hashtags via the `wrangle_and_enrich()` function call. This results in the following `pandas` DataFrame:

	username	text	created_at	lang	users	tags
0	intempestades	RT @cogautocom: "Language is how humans natur...	2021-12-13T05:23:18.000Z	en	[@cogautocom]	[]
1	pascal_bornet	RT @cogautocom: "Language is how humans natur...	2021-12-13T05:21:30.000Z	en	[@cogautocom]	[]
2	cfasfl	Dan Joldzic, CFA: Natural Language Processing ...	2021-12-13T05:08:03.000Z	en	[]	[]
3	cloohawk	New Google machine learning model is better st...	2021-12-13T05:06:41.000Z	en	[]	[#contentmarketing]
4	nisansaddspaper	Natural Language Processing for Government: Pr...	2021-12-13T04:57:30.000Z	en	[]	[]

Figure 5.10 – pandas DataFrame of Twitter search tweets

These search tweets will be perfect for creating social network visualizations as the tweets come from multiple accounts. In the top two rows of data, you may visually notice that there is a relationship between **intempestades**, **pascal_bornet**, and **cogautocom**. This would show as connected nodes if we were to visualize this network.

Converting Twitter tweets into network data

Converting social media data into network data is much easier than raw text. With Twitter, this is fortunate, as tweets tend to be quite short. This is because users frequently tag each other in their tweets for visibility and interaction, and they often associate their tweets with hashtags as well, and these associations can be used to build networks.

Using user mentions and hashtags, there are several different kinds of networks that we can create:

- *Account to Mention Networks (@ -> @)*: Useful for analyzing social networks.
- *Account to Hashtag Networks (@ -> #)*: Useful for finding communities that exist around a theme (hashtag).

- *Mention to Hashtag Network (@ -> #)*: Similar to the previous one, but linking to mentioned accounts, not the tweet account. This is also useful for finding communities.

- *Hashtag to Hashtag Networks (# -> #)*: Useful for finding related themes (hashtags) and emerging trending topics.

Additionally, you could use NER to extract additional entities from the text, but tweets are pretty short, so this may not give much useful data.

In the next few sections, you will learn how to do the first and third types of networks.

Account to Mention Network (@ -> @)

I have created a useful helper function to convert a pandas DataFrame into this Account to Mention network data:

```
def extract_user_network_data(df):
    user_network_df = df[['username', 'users', 'text']].copy()
    user_network_df = user_network_df.explode('users').dropna()
    user_network_df['users'] = user_network_df['users'].str.
replace('\@', '', regex=True)
    user_network_df.columns = ['source', 'target', 'count'] #
text data will be counted
    user_network_df = user_network_df.groupby(['source',
'target']).count()
    user_network_df.reset_index(inplace=True)
    user_network_df.sort_values(['source', 'target'],
ascending=[True, True])
    return user_network_df
```

There's quite a lot going on in this function:

1. First, we take a copy of the username, users, and text fields from the df DataFrame and use them in the user_network_df DataFrame. Each row of the users field contains a list of users, so we then "explode" the users field, creating a separate row for each user in the DataFrame. We also drop rows that do not contain any users.

2. Next, we remove all @ characters so that the data and visualization will be more readable.

3. Then, we rename all the columns in the DataFrame, in preparation for creating our graph. NetworkX's graphs expect a source and target field, and the count field can also be passed in as additional data.

4. Next, we do aggregation and count each source-target relationship in the DataFrame.

5.   Finally, we sort and return the DataFrame. We did not need to sort the DataFrame, but I tend to do this, as it can help with looking through the DataFrame or troubleshooting.

You can pass the `search_tweets` DataFrame to this function:

```
user_network_df = extract_user_network_data(df)
user_network_df.head()
```

You will get an edge list DataFrame back of user relationships. We will use this to construct and visualize our network. Look closely and you should see that there is an additional **count** field. We will use this in later chapters as a threshold for choosing which edges and nodes to show in a visualization:

	source	target	count
0	_funbot	freehipwee1	1
1	advanceml	freehipwee1	1
2	artificialbra1n	sftpmag	1
3	better_kashmiri	eslefeve	1
4	blkhwk0ps	deep__ai	1

Figure 5.11 – Account to Mention pandas DataFrame edge list

Each row in this DataFrame shows a relationship between one user (**source**) and another (**target**).

Mention to Hashtag Network (@ -> #)

I have created a useful helper function to convert a `pandas` DataFrame into **Mention** to Hashtag network data. This function is similar to the previous one, but we load users and hashtags and do not use the original account's username at all:

```
def extract_hashtag_network_data(df):
    hashtag_network_df = df[['users', 'tags', 'text']].copy()
    hashtag_network_df = hashtag_network_df.explode('users')
    hashtag_network_df = hashtag_network_df.explode('tags')
    hashtag_network_df.dropna(inplace=True)
    hashtag_network_df['users'] = hashtag_network_df['users'].
str.replace('\@', '', regex=True)
    hashtag_network_df.columns = ['source', 'target', 'count']
# text data will be counted
    hashtag_network_df = hashtag_network_df.groupby(['source',
```

```
'target']).count()
    hashtag_network_df.reset_index(inplace=True)
    hashtag_network_df.sort_values(['source', 'target'],
ascending=[True, True])
    # remove some junk that snuck in
    hashtag_network_df = hashtag_network_df[hashtag_network_
df['target'].apply(len)>2]
    return hashtag_network_df
You can pass the search_tweets DataFrame to this function.
hashtag_network_df = extract_hashtag_network_data(df)
hashtag_network_df.head()
```

You will get an edge list DataFrame back of user relationships. As shown previously, a **count** field is also returned, and we will use it in a later chapter as a threshold for choosing which nodes and edges to show:

	source	target	count
0	acdis	#ahima	2
1	acdis	#cdi	2
2	ahimaresources	#ahima	2
3	ahimaresources	#cdi	2
4	amiteshwarp	#artificialintelligence	1

Figure 5.12 – Mention to Hashtag pandas DataFrame edge list

Each row in this DataFrame shows a relationship between one user (source) and a hashtag (target).

End-to-end Twitter scraping

I hope that the preceding code and examples have shown how easy it is to use the Twitter API to scrape tweets, and I hope you can also see how easy it is to transform tweets into networks. For this chapter's final demonstration, I want you to follow a few steps:

1.   Load a pandas DataFrame containing tweets related to Network Science.

2.   Enrich the DataFrame so that it includes user mentions and hashtags as separate fields.

3.   Create Account to Mention network data.

4. Create Mention to Hashtag network data.

5. Create an Account to Mention network.

6. Create a Mention to Hashtag network.

7. Visualize the Account to Mention network.

8. Visualize the Mention to Hashtag network.

Let's do this sequentially in code, reusing the Python functions we have been using throughout this chapter.

Here is the code for the first six steps:

```
df = search_tweets(twitter_api, 'network science')
df = wrangle_and_enrich(df)
user_network_df = extract_user_network_data(df)
hashtag_network_df = extract_hashtag_network_data(df)
G_user = nx.from_pandas_edgelist(user_network_df )
G_hash = nx.from_pandas_edgelist(hashtag_network_df)
```

It is really that simple. There are a lot of moving pieces under the hood, but the more that you practice with network data, the simpler it becomes to write this kind of code.

Both of these networks are now ready for visualization:

1. I'll start with the Account to Mention network visualization. This is a social network. We can draw it like so:

    ```
    draw_graph(G_user, show_names=True, node_size=3, edge_
    width=0.5, font_size=12)
    ```

This should render a network visualization:

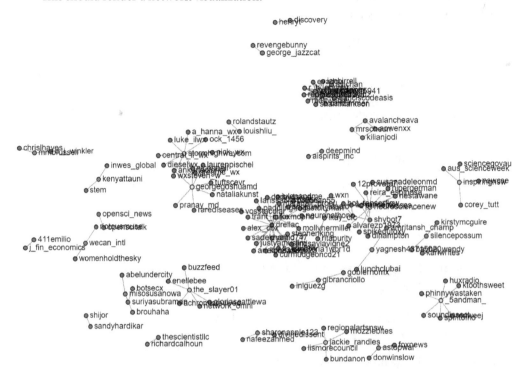

Figure 5.13 – Account to Mention social network visualization

This is a bit difficult to read since the account names overlap.

2. Let's see how the network looks without labels:

```
draw_graph(G_user, show_names=False, node_size=3, edge_
width=0.5, font_size=12)
```

This will give us a network visualization without node labels. This will allow us to see what the whole network looks like:

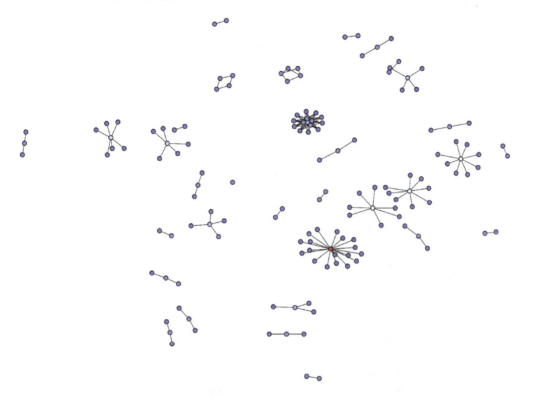

Figure 5.14 – Account to Mention social network visualization (no labels)

Wow! To me, that's beautiful and useful. I can see that there are several islands or clusters of users. If we look closer, we will be able to identify communities that exist in the data. We will do this in *Chapter 9*, which is all about Community Detection.

3. Now, let's look at the Mention to Hashtag network:

```
draw_graph(G_hash, show_names=True, node_size=3, edge_
width=0.5, font_size=12)
```

This should render a network visualization:

Figure 5.15 – Mention to Hashtag network visualization

Unlike the Account to Mention network, this one is easily readable. We can see users that are associated with various hashtags. There's no value in showing this without labels, as it is unreadable and unusable without them. This marks the end of the demonstration.

Summary

In this chapter, we covered two easier ways to scrape text data from the internet. Newspaper3k made short work of scraping news websites, returning clean text, headlines, keywords, and more. It allowed us to skip steps we'd done using BeautifulSoup and get to clean data much quicker. We used this clean text and NER to create and visualize networks. Finally, we used the Twitter Python library and V2 API to scrape tweets and connections, and we also used tweets to create and visualize networks. Between what you learned in this chapter and the previous one, you now have a lot of flexibility in scraping the web and converting text into networks so that you can explore embedded and hidden relationships.

Here is some good news: collecting and cleaning data is the most difficult part of what we are going to do, and this marks the end of data collection and most of the cleanup. After this chapter, we will mostly be having fun with networks!

In the next chapter, we will look at graph construction. We will make use of the techniques we used in this chapter to create networks for analysis and visualization.

6
Graph Construction and Cleaning

We have covered quite a lot of ground up to this point. In the previous chapters, we introduced NLP, network science, and social network analysis, and we learned how to convert raw text into network data. We even visualized a few of these networks. I hope that seeing text converted into a visualized network had the same impact on you as it did on me. The first time I attempted this, I used the book of Genesis, from the Bible, and being able to convert text from thousands of years ago into an actual interactive network took my breath away.

In the previous two chapters, we learned a few different ways to collect text data from websites and social networks on the internet and to use that text data to create networks. The good news is that I don't need to show you more ways to scrape text. You have enough options, and you should be able to use this knowledge as a foundation for other kinds of scraping.

The bad news is that it is time to get to everybody's "favorite" topic: cleaning data! In all honesty, this is my favorite part of working with network data. Cleaning takes work, but it's pretty simple. This is a good time to throw on some music, make a hot beverage, relax, and hunt for little problems to fix.

To make this chapter especially fun, we will be using the social network from *Alice's Adventures in Wonderland*. I have created this network using the process described in previous chapters. As we have gone over the steps a few times now, I'm going to skip explaining how to convert text into entities, entities into network data, and network data into graphs. The raw network data has been pushed to my GitHub, and we'll use that for this chapter.

We will be covering the following topics in this chapter:

- Creating a graph from an edge list
- Listing nodes
- Removing nodes

- Quick visual inspection
- Renaming nodes
- Removing edges
- Persisting the network
- Simulating an attack

Technical requirements

In this chapter, we will be using the NetworkX and pandas Python libraries. We will also import an NLTK tokenizer. By now, all of these libraries should be installed, so they should be ready for your use.

All the code for this chapter is available in this book's GitHub repository at https://github.com/PacktPublishing/Network-Science-with-Python.

Creating a graph from an edge list

We are going to be using this file as our original edge list: https://raw.githubusercontent.com/itsgorain/datasets/main/networks/alice/edgelist_alice_original.csv. Let's take a look:

1. Before we can create our graph, we must import the two libraries we will be working with: pandas and networkx. We use pandas to read the edge list into a DataFrame, and we pass that DataFrame to networkx to create a graph. You can import both like so:

```
import pandas as pd
import networkx as nx
```

2. With the libraries imported, let's use pandas to read the CSV file into a DataFrame and then display it, as shown in the following code block:

```
data = 'https://raw.githubusercontent.com/itsgorain/
datasets/main/networks/alice/edgelist_alice_original.csv'
network_df = pd.read_csv(data)
network_df.head()
```

If you run this in a Jupyter notebook, you should see the following DataFrame:

	source	target
0	Rabbit	Alice
1	Longitude	Alice
2	New Zealand	Ma'am
3	New Zealand	Australia
4	Fender	Alice

Figure 6.1 – pandas DataFrame of the Alice in Wonderland edge list

Before we move on, I want to say that if you can represent any two things as having a relationship, you can use that as network data. In our DataFrame, the source and target entities are people, places, and organizations, as we configured in our **Named-Entity Recognition (NER)** work, but you could also make networks of the following:

- Ingredients and dishes

- Students and teachers

- Planets and star systems

I could go on and on forever. Realizing just how prevalent networks are around us and then seeing them in everything: that's the precursor to unlocking the power of network analysis. This is about understanding the relationships between things. I am interested in literature and security, and most of my network analysis has to do with the overlap between human language and security. You may have other interests, so you will find more use in other types of networks. Try to take from this book and use it to inspire you toward new ways of researching topics of interest.

Now that that's out of the way, we have our edge list DataFrame, so let's convert this into a graph. In its simplest form, this is as easy as the following:

```
G = nx.from_pandas_edgelist(network_df)
```

Really? That's it? Yes, that's it. When I first learned that it was this easy to go from a `pandas` DataFrame to a usable graph, I was instantly hooked. It's this easy. But it is this easy for a few reasons:

- First, this `networkx` function expects **source** and **target** columns in the DataFrame. As our `.csv` file came with those columns, we didn't need to rename any columns or specify to the function which columns were our sources and targets.

- Second, we aren't specifying what kind of graph to use, so `networkx` defaults to `nx.Graph()`. This is the simplest form of graph, allowing only a single edge between nodes, and not including directionality.

In a notebook, if we were to inspect G, we would see the following:

```
G
<networkx.classes.graph.Graph at 0x255e00fd5c8>
```

This verifies that we are working with the default graph type of `Graph`.

There are several ways to load network data into `networkx`, but I prefer to use edge list. At their simplest, edge lists are tabular data with two fields: *source* and *target*. Because of this simplicity, they can easily be stored as raw text or in databases. You don't need a fancy graph database to store an edge list.

Others prefer to use an adjacency matrix when working with network data. Adjacency matrices cannot be stored easily in a database, nor do they scale out well. Use whatever you prefer, but edge lists are very easy to work with, so I recommend learning to use, create, and store them.

Types of graphs

NetworkX offers four different types of graphs:

- **Graph**
- **DiGraph**
- **MultiGraph**
- **MultiDiGraph**

You can learn more about them here: `https://networkx.org/documentation/stable/reference/classes/`. I will give a short overview and my thoughts on each type of graph as to when it is useful.

Graph

Graph is the default and simplest form of a graph that NetworkX provides. In a simple graph, nodes can have only a single edge between them and another node. If your edge list contains multiple edges between a source and a target, it'll be reduced to a single edge. This isn't always a bad thing. There are approaches to reduce the complexity of networks, and one approach is to aggregate the data – for instance, counting the number of edges that exist between two nodes and keeping that value as a weighted count, rather than the edge list being the following:

source, target

Instead, it would be like so:

source, target, edge_count

In that form, a graph still works well, because multiple edges have been reduced to a single edge, and the number of edges that existed has been reduced to a count. This is a very good way of simplifying network data while keeping all of the information.

For most of my work, a graph works fine. If I've decided against a default graph, it's because I needed directionality, so I chose a `DiGraph` instead.

Creating a default graph network can be done with the following code:

```
G = nx.from_pandas_edgelist(network_df)
```

DiGraph

A `DiGraph` is similar to a graph, with the main difference being that it is directed. DiGraph stands for **directed graph**. Just as with a graph, each node can only have one edge between itself and another node. Most of what I said about aggregation still applies, but you may need to handle self-loops if you run into them.

These are very useful when you need to understand the directionality and flow of information. It isn't always enough to know that a relationship exists between two things. It is often most important to understand the directionality of the influence, and how information spreads.

For instance, let's say we have four people named Sarah, Chris, Mark, and John. Sarah writes a lot and shares her ideas with her friend Chris. Chris is a bit of an influencer and shares information that he receives from Sarah (and others) with his following, which includes Mark and John. In this situation, the data flows like this:

Sarah -> Chris -> (Mark and John)

In this data flow, Sarah is an important person, because she is the originator of a brand-new piece of information.

Chris is also an important person because information flows through him to many other people. We will learn about how to capture this kind of importance in a later chapter, when we discuss betweenness centrality.

Finally, Mark and John are the receivers of this information.

If this were not a directed graph, we could not visibly tell who created the information or who was the final receiver of the information. This directionality allows us to go to the origins and also follow the information flow.

Directed graphs are also useful in mapping out production data flows that take place on servers and databases in production. When processes are mapped out this way, if something stops working, you can step backward until you discover what is broken. Using this approach, I have been able to troubleshoot problems in minutes that previously took days.

Creating a directed graph is as simple as this:

```
G = nx.from_pandas_edgelist(network_df, create_using=nx.
DiGraph)
```

If you inspect G, you should see the following:

```
<networkx.classes.digraph.DiGraph at 0x255e00fcb08>
```

MultiGraph

A `MultiGraph` can have multiple edges between any two nodes. A MultiGraph does not retain any context of directionality. To be truthful, I don't use `MultiGraphs`. I prefer to aggregate the multiple edges down to a count and use either Graph or `DiGraph`. However, if you want to create a MultiGraph, you can do so with the following:

```
G = nx.from_pandas_edgelist(network_df, create_using =
nx.MultiGraph)
```

If you inspect G, you will see this:

```
<networkx.classes.multigraph.MultiGraph at 0x255db7afa88>
```

MultiDiGraph

A `MultiDiGraph` can have multiple edges between any two nodes, and these graphs also convey the directionality of each edge. I do not use `MultiDiGraphs` as I prefer to aggregate multiple edges down to a count and then use either Graph or `DiGraph`. If you want to create a `MultiDiGraph`, you can do so with the following:

```
G = nx.from_pandas_edgelist(network_df, create_using =
nx.MultiDiGraph)
```

If you inspect G, you should see the following:

```
<networkx.classes.multidigraph.MultiDiGraph at 0x255e0105688>
```

Summarizing graphs

To make sure we've got everything down, let's go back over these graphs:

1. Let's recreate our graph using a default graph:

    ```
    G = nx.from_pandas_edgelist(network_df)
    ```

Great. We've loaded all of that data into G. This is a tiny network, so it loads instantly in a Jupyter notebook, and I imagine it will load quickly for you as well. With how fast of an operation that is, I'm often left wanting more, like, "That's it? I did all that work to create all of that data, and that's it?" Well, yes.

2. However, there is one function that is useful for getting a quick overview of a graph:

    ```
    print(nx.info(G))
    ```

    If we run that, we'll see this:

    ```
    Graph with 68 nodes and 68 edges
    ```

Neat. This is a tiny, simple network. With so few nodes and edges, this should visualize nicely enough to assist with cleanup.

There are other ways to quickly inspect a graph, but this is the simplest way. Now, let's look at the cleanup; we'll learn more about analyzing networks in later chapters.

Listing nodes

The first thing I tend to do after constructing a network from text is to list the nodes that have been added to the network. This allows me to take a quick peek at the node names so that I can gauge the amount of cleanup I will need to do to remove and rename nodes. During our entity extraction, we had the opportunity to clean the entity output. The entity data is used to create the network data that is used to create the graph itself, so there are multiple steps during which cleanup and optimization are possible, and the more that you do upstream, the less that you have to do later.

However, it is still important to take a look at the node names, to identify any strangeness that still managed to find a way into the network:

1. The simplest way to get a node list is to run the following networkx command:

    ```
    G.nodes
    ```

    This will give you a NodeView:

    ```
    NodeView(('Rabbit', 'Alice', 'Longitude', 'New Zealand',
    "Ma'am", 'Australia', 'Fender', 'Ada', 'Mabel', 'Paris',
    'Rome', 'London', 'Improve', 'Nile', 'William the
    Conqueror', 'Mouse', 'Lory', 'Eaglet', 'Northumbria',
    'Edwin', 'Morcar', 'Stigand', 'Mercia', 'Canterbury', 'â\
    x80\x98it', 'William', 'Edgar Atheling', "â\x80\x98I'll",
    'Said', 'Crab', 'Dinah', 'the White Rabbit', 'Bill',
    'The Rabbit Sends', 'Mary Ann', 'Pat', 'Caterpillar',
    'CHAPTER V.', 'William_', 'Pigeon', 'Fish-Footman',
    'Duchess', 'Cheshire', 'Hare', 'Dormouse', 'Hatter',
    ```

```
'Time', 'Tillie', 'Elsie', 'Lacie', 'Treacle', 'Kings',
'Queens', 'Cat', 'Cheshire Cat', 'Somebody', 'Mystery',
'Seaography', 'Lobster Quadrille', 'France', 'England',
'â\x80\x98Keep', 'garden_.', 'Hm', 'Soup', 'Beautiful',
'Gryphon', 'Lizard'))
```

2. That's readable, but it could be a little easier on the eyes. This function will clean it up a bit:

```
def show_nodes(G):
    nodes = sorted(list(G.nodes()))
    return ', '.join(nodes)
```

This can be run as follows:

```
show_nodes(G)
```

This outputs a cleaner node list:

```
"Ada, Alice, Australia, Beautiful, Bill, CHAPTER V.,
Canterbury, Cat, Caterpillar, Cheshire, Cheshire Cat,
Crab, Dinah, Dormouse, Duchess, Eaglet, Edgar Atheling,
Edwin, Elsie, England, Fender, Fish-Footman, France,
Gryphon, Hare, Hatter, Hm, Improve, Kings, Lacie, Lizard,
Lobster Quadrille, London, Longitude, Lory, Ma'am, Mabel,
Mary Ann, Mercia, Morcar, Mouse, Mystery, New Zealand,
Nile, Northumbria, Paris, Pat, Pigeon, Queens, Rabbit,
Rome, Said, Seaography, Somebody, Soup, Stigand, The
Rabbit Sends, Tillie, Time, Treacle, William, William the
Conqueror, William_, garden_., the White Rabbit, â\x80\
x98I'll, â\x80\x98Keep, â\x80\x98it"
```

We now have a clean list of nodes that exist in the *Alice's Adventures in Wonderland* social network. Immediately, my eyes are drawn to the last three nodes. These don't even look like names. We're going to remove them. I can also see that CHAPTER V., Soup, and a few other non-entity nodes were added. This is a common problem when using NLP for Part-of-Speech Tagging (pos_tagging) or **NER**. Both of these approaches frequently make mistakes on words where the first letter of a word is capitalized.

We have some work to do. We will remove the nodes that were added by mistake, and we will rename a few of the nodes so that they reference the *White Rabbit*.

When inspecting graphs, I list nodes, not edges. You can list edges with the following:

```
G.edges
```

This will give you an `EdgeView`, like this:

```
EdgeView([('Rabbit', 'Alice'), ('Rabbit', 'Mary Ann'),
('Rabbit', 'Pat'), ('Rabbit', 'Dinah'), ('Alice', 'Longitude'),
('Alice', 'Fender'), ('Alice', 'Mabel'), ('Alice', 'William the
Conqueror'), ('Alice', 'Mouse'), ('Alice', 'Lory'), ('Alice',
'Mary Ann'), ('Alice', 'Dinah'), ('Alice', 'Bill'), ('Alice',
'Caterpillar'), ('Alice', 'Pigeon'), ('Alice', 'Fish-Footman'),
('Alice', 'Duchess'), ('Alice', 'Hare'), ('Alice', 'Dormouse'),
('Alice', 'Hatter'), ('Alice', 'Kings'), ('Alice', 'Cat'),
('Alice', 'Cheshire Cat'), ('Alice', 'Somebody'), ('Alice',
'Lobster Quadrille'), ('Alice', 'â\x80\x98Keep'), ('Alice',
'garden_.'), ('Alice', 'Hm'), ('Alice', 'Soup'), ('Alice',
'the White Rabbit'), ('New Zealand', "Ma'am"), ('New Zealand',
'Australia'), ('Ada', 'Mabel'), ('Paris', 'Rome'), ('Paris',
'London'), ('Improve', 'Nile'), ('Mouse', 'â\x80\x98it'),
('Mouse', 'William'), ('Lory', 'Eaglet'), ('Lory', 'Crab'),
('Lory', 'Dinah'), ('Northumbria', 'Edwin'), ('Northumbria',
'Morcar'), ('Morcar', 'Stigand'), ('Morcar', 'Mercia'),
('Morcar', 'Canterbury'), ('William', 'Edgar Atheling'), ("â\
x80\x98I'll", 'Said'), ('the White Rabbit', 'Bill'), ('the
White Rabbit', 'The Rabbit Sends'), ('Caterpillar', 'CHAPTER
V.'), ('Caterpillar', 'William_'), ('Duchess', 'Cheshire'),
('Duchess', 'Cat'), ('Duchess', 'Lizard'), ('Hare', 'Hatter'),
('Hare', 'Lizard'), ('Dormouse', 'Hatter'), ('Dormouse',
'Tillie'), ('Dormouse', 'Elsie'), ('Dormouse', 'Lacie'),
('Dormouse', 'Treacle'), ('Hatter', 'Time'), ('Kings',
'Queens'), ('Mystery', 'Seaography'), ('France', 'England'),
('Soup', 'Beautiful'), ('Soup', 'Gryphon')])
```

I don't typically list edges, because when I remove or rename nodes, the edges will be corrected. Edges to nodes that have been removed will be removed as well. Edges to nodes that have been renamed will be connected to the renamed node. `EdgeView` is also more confusing to look at.

With our clean nodelist, here is our plan of attack:

1. Remove the bad nodes.
2. Rename the *White Rabbit* nodes.
3. Add any missing nodes that I am aware of.
4. Add any missing edges that I can identify.

Let's proceed with the first of those steps.

Removing nodes

The next thing we will do is remove nodes that have made it into the network by mistake, usually as a result of false positives from `pos_tagging` or NER. You may see me refer to these nodes as "bad" nodes. I could as easily refer to them as "unwanted" nodes, but the point is that these are nodes that do not belong and should be removed. For simplicity, I call them bad nodes.

One reason to remove nodes is to clean a network so that it closely matches reality or the reality described in a piece of text. However, removing nodes can also be useful, for simulating an attack. We could, for instance, remove key characters from the *Alice in Wonderland* social network, to simulate what the outcome would be if the Queen of Hearts had gotten her wish of executing several characters. We will do that in this chapter.

Simulating an attack is also useful for bolstering defenses. If a node is a single point of failure and if its removal would be catastrophic to a network, you can potentially add nodes in certain positions so that if the critical node were removed, the network would remain intact, and information flow would be undisrupted:

- In `networkx`, there are two different ways to remove nodes: one at a time, or several at once. You can remove a single node like this:

  ```
  G.remove_node('â\x80\x98it')
  ```

- You can remove several nodes at once like this:

  ```
  drop_nodes = ['Beautiful', 'CHAPTER V.', 'Hm', 'Improve',
  'Longitude', 'Ma\'am', 'Mystery', 'Said', 'Seaography',
  'Somebody', 'Soup', 'Time', 'garden_.', 'â\x80\x98I\'ll',
  'â\x80\x98Keep']
  G.remove_nodes_from(drop_nodes)
  ```

I prefer the second approach because it can also be used to remove a single node if the `drop_nodes` variable only contains a single node name. You can simply keep expanding `drop_nodes` until you have all bad entities listed, and then you can keep refreshing the list of remaining nodes. Now that we've removed some nodes, let's see which entities remain:

```
show_nodes(G)
'Ada, Alice, Australia, Bill, Canterbury, Cat, Caterpillar,
Cheshire, Cheshire Cat, Crab, Dinah, Dormouse, Duchess, Eaglet,
Edgar Atheling, Edwin, Elsie, England, Fender, Fish-Footman,
France, Gryphon, Hare, Hatter, Kings, Lacie, Lizard, Lobster
Quadrille, London, Lory, Mabel, Mary Ann, Mercia, Morcar,
Mouse, New Zealand, Nile, Northumbria, Paris, Pat, Pigeon,
```

```
Queens, Rabbit, Rome, Stigand, The Rabbit Sends, Tillie,
Treacle, William, William the Conqueror, William_, the White
Rabbit'
```

This is already looking much cleaner. Next, we will further clean the network by renaming and combining certain nodes, especially the nodes related to the *White Rabbit*.

Quick visual inspection

Before moving on to more cleaning, let's do a quick visual inspection of the network. We will reuse the draw_graph function we have been using throughout this book:

```
draw_graph(G, show_names=True, node_size=5, edge_width=1)
```

This outputs the following network:

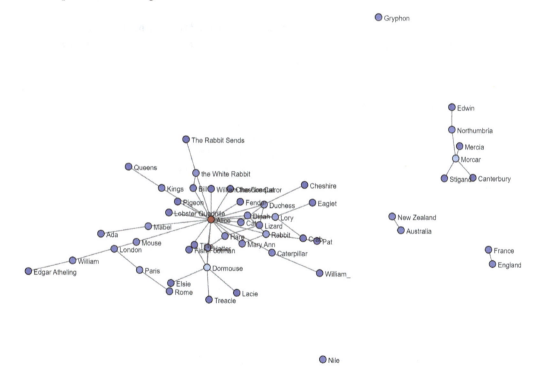

Figure 6.2 – Quick visual inspection network

OK, what do we see? I can see that there is one large cluster of connected entities. This is the primary component of the Alice in Wonderland network.

What else do we see? **Alice** is the most central node in the primary component. That makes sense, as she is the main character in the story. Thinking about the main characters, I see many names that I know, such as **Dormouse**, **Cheshire Cat**, and **White Rabbit**. What is interesting to me, though, is that not only are they shown but I can also begin to see which characters are most important to the story based on the number of entities connected to them. However, I also see that the Queen and King of Hearts are missing, which is disappointing. NER failed to recognize them as entities. From what I have seen, NER struggles with fantasy and ancient names, due to it being trained on more realistic data. It would struggle less with real names. We will manually add several members of the queen's court, including the king and queen.

I can also see a few strange nodes that seem to be part of the story but they aren't connected to the primary component. Why is **Gryphon** disconnected from everything? Who does **Gryphon** know? We should look for those relationships in the story text. We will manually add the edges.

Finally, I see nodes that have to do with places on Earth, such as **Nile**, **France**, **England**, **New Zealand**, and **Australia**. We could keep these, as they are technically a part of the story, but I'm going to remove them so that we can focus more on the social network of character relationships that exist in Wonderland. We will remove these.

Let's start by removing the non-Wonderland nodes:

```
drop_nodes = ['New Zealand', 'Australia', 'France', 'England',
'London', 'Paris', 'Rome', 'Nile', 'William_', 'Treacle',
'Fender', 'Canterbury', 'Edwin', 'Mercia', 'Morcar',
'Northumbria', 'Stigand']
G.remove_nodes_from(drop_nodes)
```

Now, let's visualize the network again:

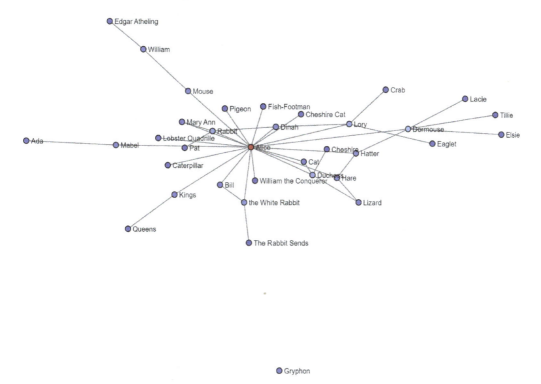

Figure 6.3 – Quick visual inspection network (cleaned)

That looks a lot better. We still have **Gryphon** floating around like an island, but we'll take care of that soon. Still, where in the world is the Queen of Hearts? I wrote a helper function to help with that investigation:

```
from nltk.tokenize import sent_tokenize
def search_text(text, search_string):
    sentences = sent_tokenize(text)
    for sentence in sentences:
        if search_string in sentence.lower():
            print(sentence)
            print()
```

With this function, we can pass in any text and any search string, and it'll print out any sentences that contain the search string. This will help us find entities and relationships that NER failed to find. I am using NLTK's sentence tokenizer rather than spaCy because this is faster and easier to get the results I need right now. Sometimes, NLTK is the faster and simpler approach, but not in this case.

> **Note**
>
> To run the following code, you will need to load the text variable using one of the approaches from *Chapter 4* or *Chapter 5*. We have shown multiple approaches to loading text for Alice in Wonderland.

Let's look for text related to the queen:

```
search_text(text, 'queen')
```

Here are some of the results:

```
An invitation from the  Queen to play croquet.""
The Frog-Footman repeated, in the same solemn  tone, only
changing the order of the words a little, "From the Queen.
"I must go and get ready to play  croquet with the Queen,"" and
she hurried out of the room.
"Do you play croquet with the  Queen to-day?""
"We  quarrelled last March just before _he_ went mad, you know
"" (pointing  with his tea spoon at the March Hare,) " it was
at the great concert  given by the Queen of Hearts, and I had
to sing    âTwinkle, twinkle, little bat!
"Well, I'd hardly finished the first verse,"" said the Hatter,
"when the  Queen jumped up and bawled out, âHe's murdering the
time!
The Queen's Croquet-Ground      A large rose-tree stood near
the entrance of the garden: the roses  growing on it were
white, but there were three gardeners at it, busily  painting
them red.
"I heard the Queen say only  yesterday you deserved to be
beheaded!""
```

Running this function will give many more results than this – these are just a few. But we can already see that Queen of Hearts knows Frog-Footman, and that Frog-Footman is in our network, so we should add Queen of Hearts and other missing characters and place an edge between the characters that they interact with.

Adding nodes

We need to add nodes that are missing. As *Alice in Wonderland* is a fantasy story, and NER models tend to be trained with more modern and realistic text, the NER struggled to identify some important entities, including the Queen of Hearts. There are a few lessons here:

- First, don't blindly trust models, ever. The data that they were trained on will have an impact on what they do and don't do very well.

- Second, domain knowledge is very important. If I did not know the story of Alice in Wonderland, I might not even have noticed that the royalty was missing.

- Finally, even with flaws, NER and these approaches will do *most* of the work in converting text into networks, but your domain knowledge and critical thinking will lead to the best results.

Just as with removing nodes, networkx has two methods for adding nodes: one at a time, or several at once:

- We can add just 'Queen of Hearts':

  ```
  G.add_node('Queen of Hearts')
  ```

- Alternatively, we could add the missing nodes all at once:

  ```
  add_nodes = ['Queen of Hearts', 'Frog-Footman', 'March
  Hare', 'Mad Hatter', 'Card Gardener #1', 'Card Gardener
  #2', 'Card Gardener #3', 'King of Hearts', 'Knave of
  Hearts', 'Mock Turtle']
  G.add_nodes_from(add_nodes)
  ```

Again, I prefer the bulk approach, as I can just keep extending the add_nodes list until I am satisfied with the results. If we visualize the network now, these added nodes will appear as islands, because we have not created edges between them and other nodes:

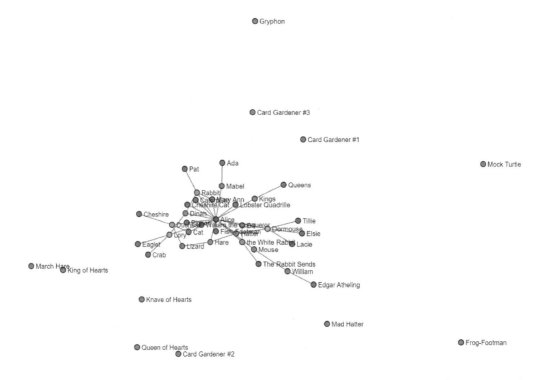

Figure 6.4 – Network with missing nodes added

This looks good. Next, let's add those missing edges.

Adding edges

We used the search_text function to identify not only the missing characters but also the missing relationships between those characters. The approach taken was as follows:

1. Figure out who the Queen of Hearts knows; take notes as these are missing edges.

2. Add the Queen of Hearts and any other missing nodes.

3. Figure out who each missing character knows; take notes as these are missing edges.

This involved doing a bunch of lookups with the `search_text` function and then keeping track of relationships as comments in my Jupyter notebook. In the end, it looked like this:

```
# Alice -> Mock Turtle
# King of Hearts -> Alice
# King of Hearts -> Card Gardener #1
# King of Hearts -> Card Gardener #2
# King of Hearts -> Card Gardener #3
# King of Hearts -> Dormouse
# King of Hearts -> Frog-Footman
# King of Hearts -> Kings
# King of Hearts -> Lizard
# King of Hearts -> Mad Hatter
# King of Hearts -> March Hare
# King of Hearts -> Mock Turtle
# King of Hearts -> Queen of Hearts
# King of Hearts -> Queens
# King of Hearts -> White Rabbit
# Knave of Hearts -> King of Hearts
# Knave of Hearts -> Queen of Hearts
# Queen of Hearts -> Alice
# Queen of Hearts -> Card Gardener #1
# Queen of Hearts -> Card Gardener #2
# Queen of Hearts -> Card Gardener #3
# Queen of Hearts -> Dormouse
# Queen of Hearts -> Frog-Footman
# Queen of Hearts -> Kings
# Queen of Hearts -> Lizard
# Queen of Hearts -> Mad Hatter
# Queen of Hearts -> March Hare
# Queen of Hearts -> Mock Turtle
# Queen of Hearts -> Queens
# Queen of Hearts -> White Rabbit
```

Figure 6.5 – Identified missing edges

These are identified edges that we need to add. We are likely missing some, but this is enough for our purposes:

- We can add an edge, one at a time:

```
G.add_edge('Frog-Footman', 'Queen of Hearts')
```

- Alternatively, we can add several at once. I prefer the bulk approach, again. To do the bulk approach, we will use a list of tuples to describe the edges:

```
add_edges = [('Alice', 'Mock Turtle'), ('King of Hearts',
'Alice'), ('King of Hearts', 'Card Gardener #1'),
            ('King of Hearts', 'Card Gardener #2'),
('King of Hearts', 'Card Gardener #3'),
            ('King of Hearts', 'Dormouse'), ('King of
Hearts', 'Frog-Footman'), ('King of Hearts', 'Kings'),
            ('King of Hearts', 'Lizard'), ('King of
Hearts', 'Mad Hatter'), ('King of Hearts', 'March Hare'),
            ('King of Hearts', 'Mock Turtle'), ('King
of Hearts', 'Queen of Hearts'), ('King of Hearts',
'Queens'),
            ('King of Hearts', 'White Rabbit'), ('Knave
of Hearts', 'King of Hearts'),
            ('Knave of Hearts', 'Queen of Hearts'),


            ('Queen of Hearts', 'Alice'), ('Queen of
Hearts', 'Card Gardener #1'),
            ('Queen of Hearts', 'Card Gardener #2'),
('Queen of Hearts', 'Card Gardener #3'),
            ('Queen of Hearts', 'Dormouse'), ('Queen of
Hearts', 'Frog-Footman'), ('Queen of Hearts', 'Kings'),
            ('Queen of Hearts', 'Lizard'), ('Queen
of Hearts', 'Mad Hatter'), ('Queen of Hearts', 'March
Hare'),
            ('Queen of Hearts', 'Mock Turtle'), ('Queen
of Hearts', 'Queens'), ('Queen of Hearts', 'White
Rabbit')]

G.add_edges_from(add_edges)
```

How does our network look now?

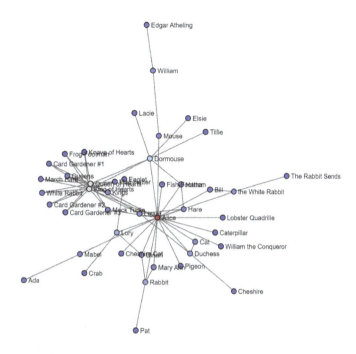

Figure 6.6 – Network with missing edges added

This is looking a lot better, and we now have the queen's court in place. However, Gryphon is still an island, so let's do a lookup to see what relationship or relationships are missing:

```
search_text(text, 'gryphon')
```

This gives us some text to look at, and I used that to identify missing edges. Let's add them:

```
add_edges = [('Gryphon', 'Alice'), ('Gryphon', 'Queen of
Hearts'), ('Gryphon', 'Mock Turtle')]


G.add_edges_from(add_edges)
```

Now, let's visualize the network one more time:

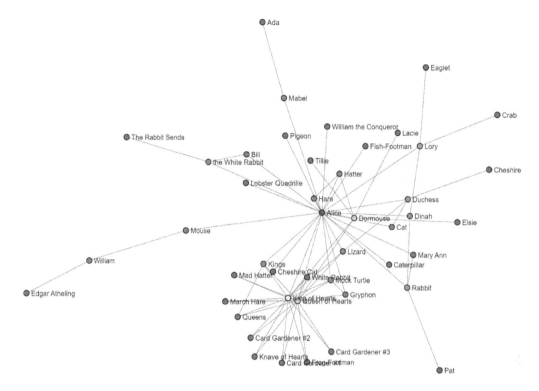

Figure 6.7 – Network with missing edges added (final)

Ah! It's such a wonderful feeling when a disconnected network is finally connected and all islands/isolates are taken care of. This is clean and readable. We have successfully removed junk nodes, added missing nodes, and connected the missing nodes to nodes that they should share an edge with! We can move on!

Renaming nodes

This network looks good enough that we might be tempted to just call it a day on our cleaning efforts. However, there is a bit more that we need to do, especially for the White Rabbit, but also for a few other characters. I can see three nodes having to do with the White Rabbit:

- `the White Rabbit`
- `Rabbit`
- `The Rabbit Sends`

If we rename all three of these nodes White Rabbit, then they will be combined into a single node, and their edges will also be correctly connected. There are a few other nodes that should be renamed as well. Here is how to rename nodes:

```
relabel_mapping = {'Cheshire':'Cheshire Cat', 'Hatter':'Mad
Hatter', 'Rabbit':'White Rabbit','William':'Father William',
'the White Rabbit':'White Rabbit', 'The Rabbit Sends':'White
Rabbit', 'Bill':'Lizard Bill', 'Lizard':'Lizard Bill',
'Cat':'Cheshire Cat', 'Hare':'March Hare'}
G = nx.relabel_nodes(G, relabel_mapping)
```

We pass in a Python dictionary containing nodes and what we want them relabeled as. For instance, we are changing Cheshire to Cheshire Cat, and Hatter to Mad Hatter.

How do our nodes look now?

```
show_nodes(G)
'Ada, Alice, Card Gardener #1, Card Gardener #2, Card Gardener
#3, Caterpillar, Cheshire Cat, Crab, Dinah, Dormouse, Duchess,
Eaglet, Edgar Atheling, Elsie, Father William, Fish-Footman,
Frog-Footman, Gryphon, King of Hearts, Kings, Knave of Hearts,
Lacie, Lizard Bill, Lobster Quadrille, Lory, Mabel, Mad Hatter,
March Hare, Mary Ann, Mock Turtle, Mouse, Pat, Pigeon, Queen of
Hearts, Queens, Tillie, White Rabbit, William the Conqueror'
```

Nice. That looks perfect. How does our network look, visually?

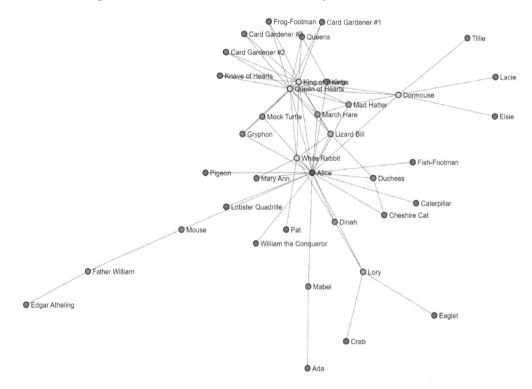

Figure 6.8 – Network with renamed nodes

Perfect. **White Rabbit** has been placed correctly, and the node color and placement show it as a central character, right next to **Alice**, and not far from **Queen of Hearts** and **King of Hearts**.

Removing edges

There will likely be times when you will need to remove edges. This can be useful, not just for cleaning networks but also for simulating attacks, or for identifying cliques and communities. For instance, I often use what is called **minimum cuts** or **minimum edge cuts** to find the fewest number of edges that will split a network into two pieces. I use this for community detection, and also to spot emerging trends on social media.

With the *Alice in Wonderland* network, there are no edges that we need to remove, so I will first show you how to remove some edges, and then I'll show you how to put them back:

- You can remove edges one at a time:

```
G.remove_edge('Dormouse', 'Tillie')
```

- Alternatively, you can remove several at a time:

```
drop_edges = [('Dormouse', 'Tillie'), ('Dormouse',
'Elsie'), ('Dormouse', 'Lacie')]
G.remove_edges_from(drop_edges)
```

How does this look when visualized?

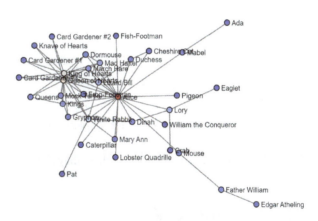

Figure 6.9 – Network with edges removed

This looks exactly as it should. If we had removed the nodes for **Elsie**, **Tillie**, and **Lacie** instead of their edges, then the nodes as well as the edges would have been removed. Instead, we have removed the edges, which is a bit like cutting a piece of string. The three nodes are now islands, isolates, connected to nothing.

Let's put them back:

```
add_edges = [('Dormouse', 'Elsie'), ('Dormouse', 'Lacie'),
('Dormouse', 'Tillie')]
G.add_edges_from(add_edges)
```

How does the network look now?

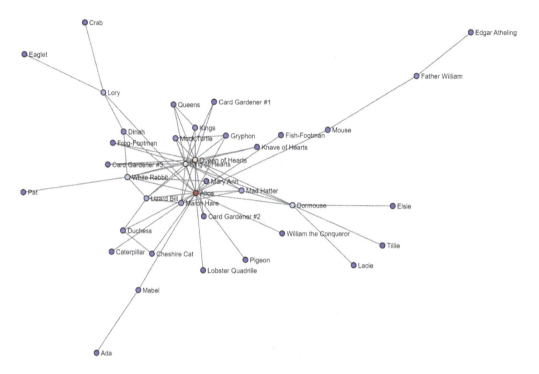

Figure 6.10 – Network with edges added

Perfect. **Elsie**, **Tillie**, and **Lacie** are right back where they should be, connected to **Dormouse**. With that, I think this network is perfect for our uses.

Persisting the network

I want to persist this network so that we can use it in later chapters without having to go through all of this work again. We will use this network quite a lot in this book:

```
outfile = r'C:\blah\blah\blah\networks\alice\edgelist_alice_
cleaned.csv'
final_network_df = nx.to_pandas_edgelist(G)
final_network_df.to_csv(outfile, header=True, index=False)
```

I'm using Microsoft Windows. Your outfile path may look different.

Simulating an attack

We already did an end-to-end workflow of converting a rough network edge list into a network, cleaning the network, and then persisting the cleaned network's edge list, so for the remainder of this chapter, let's do a simulation.

In most networks, some nodes serve as key hubs. These nodes reveal themselves if you look for the number of degrees (edges) that a node has, or by checking PageRank or various centrality metrics for nodes. We will use these approaches in a later chapter to identify important notes. For now, we have domain knowledge that we can use. Those of us who know this story can likely name by heart several of the important protagonists of the story: Alice, Mad Hatter, Cheshire Cat, and so on. And those of us who are familiar with the story are also likely very aware of the Queen of Hearts repeatedly shouting "OFF WITH THEIR HEADS!"

In a network, if you remove the most connected and important nodes, what often happens looks a lot like the scene in *Star Wars*, where the Death Star explodes. All at once, many nodes are transformed into isolates, their edges destroyed along with the central node that was removed. This is catastrophic to a network, and the information flow is disrupted. Can you imagine the real-world impact of what happens when key nodes are removed from a network? Your internet goes down. Your power goes out. Supply chains are disrupted. Grocery stores are not stocked, and on and on and on. Understanding networks and simulating disruption can give ideas on how to bolster the supply chain and information flow. That is the point of this exercise.

But we are going to have fun. We are just going to give the Queen of Hearts one huge win. We are going to let her execute four of the main characters in the story and see what happens:

1. First, let's execute them:

    ```
    drop_nodes = ['Alice', 'Dormouse', 'White Rabbit', 'Mad
    Hatter']
    G.remove_nodes_from(drop_nodes)
    ```

    We decided that the Queen of Hearts has successfully executed Alice, Dormouse, White Rabbit, and Mad Hatter. It would have been a terrible story if this had happened, but we're going to play it out. I am choosing these four because I know that they are key characters in the story. Their removal from the network should shatter it, which is what I want to demonstrate.

2. After removing just four of the key nodes in this network, what does the rest of the network look like? What are the consequences?

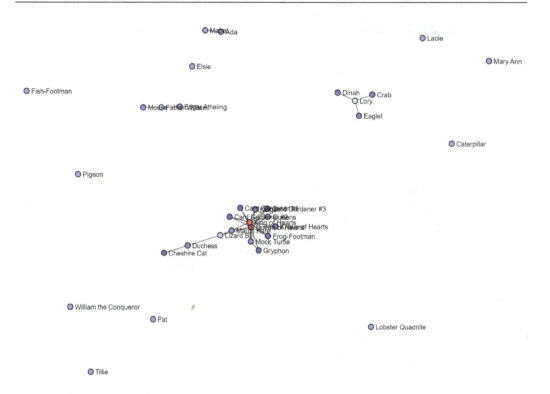

Figure 6.11 – Shattered Alice network

Disaster. We can see several nodes that were made into isolates. We still have one primary component in the center, and we have two other smaller components with two to four nodes. But in general, the network has been shattered, and the information flow has been disrupted. New relationships will need to be built. New hierarchies will need to be established. The queen's court has become dominant, just by removing four nodes.

3. Let's look closer at the primary component:

```
components = list(nx.connected_components(G))
main_component = components[4]
G_sub = G.subgraph(main_component)
draw_graph(G_sub, show_names=True, node_size=4, edge_
width = 0.5)
```

There are a few things to understand in this code. First, nx.connected_components(G) has converted the graph into a list of connected components. One of the components will be the primary component, but it is not necessarily the first one on the list. After some investigation, we will find that the fourth component was the primary component, so let's set that as main_ component and then visualize the subgraph of that component. This is what we see:

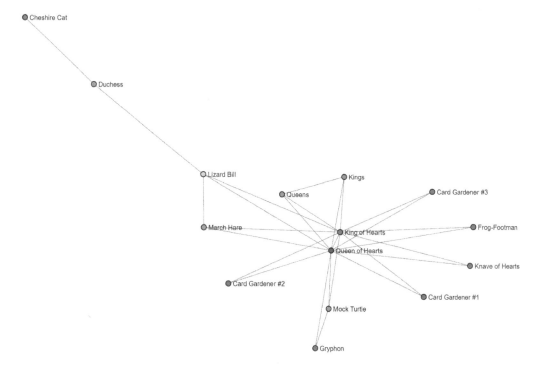

Figure 6.12 – Queen's court subgraph

The queen's court is intact and contains characters who were unfortunate enough to be trapped in the network before the executions happened.

And that's it for this chapter!

Summary

In this chapter, we took raw data, performed a series of steps to clean the network, and even carried out a very simple attack simulation.

I hope that at this point, looking at and working with networks is starting to feel more natural. The more that I work with networks, the more that I see them in everything, and they affect my understanding of the world. We are using a fantasy story for this chapter because it is of a manageable size to explain the construction, cleaning, and some simple analysis. As you learn more about networks, you will likely find that real-world networks are usually much messier, more complicated, and larger. I hope that this simple network will give you the tools and practice you need to eventually chase much more ambitious problems.

In the next chapter, we're going to have a lot of fun. Our next chapter is about analyzing whole networks. You will learn all kinds of useful things, such as how to identify the most influential nodes in a network. From here on out, we will do a lot of network analysis and visualization.

Part 3: Network Science and Social Network Analysis

In these chapters, we learn how to analyze networks and hunt for insights. We begin with a discussion on whole network analysis and gradually zoom in to the node level, to investigate egocentric networks. We then look for communities and subgroups that exist in networks. Finally, we conclude the book by showing how graph data can be useful for supervised and unsupervised machine learning.

This section includes the following chapters:

- *Chapter 7, Whole Network Analysis*
- *Chapter 8, Egocentric Network Analysis*
- *Chapter 9, Community Detection*
- *Chapter 10, Supervised Machine Learning on Network Data*
- *Chapter 11, Unsupervised Machine Learning on Network Data*

7
Whole Network Analysis

In previous chapters, we spent a lot of time covering how networks can be constructed using text and how cleanup can be done on networks. In this chapter, we are moving on to **whole network analysis**. For the sake of simplicity, I will call it **WNA**. WNA is done to get the lay of the land, to understand the denseness of a network, which nodes are most important in various ways, which communities exist, and so forth. I'm going to cover material that I have found useful, which is a bit different from what is found in most **social network analysis (SNA)** or network science books. I do applied network science every day, and my goal is to showcase some of the options that are available to allow readers to very quickly get started in network analysis.

Network science and SNA are both very rich topics, and if you find any section of this chapter especially interesting, I encourage you to do your own research to learn more. Throughout this book, I will reference certain sections of NetworkX documentation. Be aware that there are many additional non-covered capabilities on those reference pages, and it can be insightful to learn about lesser-used functions, what they do, and how they can be used.

NetworkX's online documentation shares links to journal articles, so there is plenty to read and learn.

As you read this chapter, I want you to consider the problems that you work on and try to find ways that you can use what I am describing in your own work. Once you begin working with networks, you will see that they are everywhere, and once you learn how to analyze and manipulate them, then a world of opportunity opens up.

We will cover the following topics:

- Creating baseline WNA questions
- WNA in action
- Comparing centralities
- Visualizing subgraphs
- Investigating connected components
- Understanding network layers

Technical requirements

In this chapter, we will be using the Python libraries NetworkX and pandas. Both of these libraries should be installed by now, so they should be ready for your use. If they are not installed, you can install Python libraries with the following:

```
pip install <library name>
```

For instance, to install NetworkX, you would do the following:

```
pip install networkx
```

In *Chapter 4*, we also introduced a `draw_graph()` function that uses both NetworkX and `Scikit-Network`. You will need that code anytime that we do network visualization. Keep it handy!

You can find all of the code in this chapter in the GitHub repository: `https://github.com/PacktPublishing/Network-Science-with-Python`.

Creating baseline WNA questions

I often jot down questions that I have before doing any kind of analysis. This sets the context of what I am looking for and sets up a framework for me to pursue those answers.

In doing any kind of WNA, I am interested in finding answers to each of these questions:

- How big is the network?
- How complex is the network?
- What does the network visually look like?
- What are the most important nodes in the network?
- Are there islands, or just one big continent?
- What communities can be found in the network?
- What bridges exist in the network?
- What do the layers of the network reveal?

These questions give me a start that I can use as a task list for running through network analysis. This allows me to have a disciplined approach when doing network analysis, and not just chase my own curiosity. Networks are noisy and chaotic, and this scaffolding gives me something to use to stay focused.

Revised SNA questions

In this chapter, we will be using a K-pop social network. You can learn more about this network data in *Chapter 2*.

My goal is to understand the shape of the network and how information flows between individuals and communities. I also want to be able to explore different levels of the network, as if I am peeling an onion. The core is often especially interesting.

As this is a social network, I have additional questions beyond the previous baseline questions:

- How big is the social network? What does this mean?
- How complex and interconnected is the network?
- What does the network visually look like?
- Who are the most important people and organizations in the network?
- Is there just one giant cluster in the network, or are there isolated pockets of people?
- What communities can be found in the network?
- What bridges exist in the network?
- What do the layers of the network reveal?

Social network analysis revisited

In *Chapter 2, Network Analysis*, I described the definition, origins, and uses of network science and SNA. Although these are two independent fields of study, there is so much overlap that I consider the social network to be a set of techniques that should be rolled up into network science. This is because SNA can make great use of network science tools and techniques, and network science can be made a lot more interesting by applying it to social networks. I personally do not distinguish between the two.

What *is* social network analysis? In my view, it is a different perspective on network analysis, from a social angle. Network science has to do with how networks are constructed, the properties of networks, and how networks evolve over time. In social network analysis, we are interested in getting a bit more personal. We want to know *who* the important people and organizations are that exist in a network, which individuals serve as bridges between communities, and which communities exist and why they exist.

Content analysis is where the marriage of NLP and network science is most important. NLP allows for the extraction of entities (people, places, and organizations) and predicts the sentiment of classifying text. Network science and SNA allow for understanding much more about the relationships that exist in these networks. So, with NLP and network analysis, you have both content context as well as relationship context. This is a powerful synergy, where *1 + 1 = 3*.

In this chapter, we are not going to be doing any NLP. I will explain some capabilities of network science and SNA. So, let's get started!

WNA in action

As mentioned in the previous chapter, in NetworkX, you are able to construct networks as either undirected, directed, multi-, or multi-directed graphs. In this chapter, we're going to use an undirected graph, as I want to show how certain functionality can be used to understand networks. Just know this: what I am about to show has different implications if you use one of the other types of networks. You also have more options to explore when you use directed networks, such as investigating in_degrees and out_degrees, not just degrees in general.

Loading data and creating networks

The first thing we need to do is construct our graph. We cannot analyze what we do not have:

1. You can read the K-pop edge list from my GitHub like so:

    ```
    import pandas as pd
    data = 'https://raw.githubusercontent.com/itsgorain/
    datasets/main/networks/kpop/kpop_edgelist.csv'
    df = pd.read_csv(data)
    df['source'] = df['source'].str[0:16]
    df['target'] = df['target'].str[0:16]
    df.head()
    ```

    Previewing the pandas DataFrame, we can see that there are columns for 'source' and 'target'. This is exactly what NetworkX is looking for to build a graph. If you had wanted to name the graph columns differently, NetworkX would have allowed you to specify your own source and target columns.

2. Looking at the shape of the edge list, we can see that there are 1,286 edges in the edge list:

    ```
    df.shape[0]
    1286
    ```

    Remember, an edge is a relationship between one node and another, or between one node and itself, which is known as a **self-loop**.

3. Now that we have our pandas edge list ready, we can use it to construct our undirected graph:

    ```
    import networkx as nx
    G = nx.from_pandas_edgelist(df)
    ```

```
G.remove_edges_from(nx.selfloop_edges(G))
G.remove_node('@') # remove a junk node
```

4. Finally, let's inspect G to make sure that it is an undirected NetworkX graph:

    ```
    G
    ```

    ```
    <networkx.classes.graph.Graph at 0x217dc82b4c8>
    ```

    This looks perfect, so we are ready to begin our analysis.

Network size and complexity

The first thing we are going to investigate is the network's size, shape, and overall complexity. Let me define what I mean by that:

- **Network size**: The number of nodes and the number of edges that exist in a network
- **Network complexity**: The amount of clustering and density present in the network. Clustering is the number of possible triangles that actually exist in a network, and density similarly refers to how interconnected the nodes in a network are.

NetworkX makes it very easy to find the number of nodes and edges that exist in a network. You can simply use nx.info(G), like so:

```
nx.info(G)
'Graph with 1163 nodes and 1237 edges'
```

Our network has 1,163 nodes and 1,237 edges. To put that into plain English, our K-pop social network consists of 1,163 people and organizations, and among those 1,163 people and organizations, there are 1,237 identified interactions. As this is Twitter data, an interaction, in this case, means that the two accounts were mentioned in the same tweet, meaning that they are related in some way. Going back to the importance of NLP and content analysis, we can use these identified relationships to further dig into what types of relationships these actually are. Are they collaborative relationships? Were they arguing? Did they write a research paper together? SNA will not give us that answer. We need content analysis to get to those. But this chapter is on network analysis, so let's continue.

Is this a dense network? Internet social networks tend to be sparse, not dense, unless you are analyzing a tight-knit group of people.

Let's see what the clustering and density of the network look like:

1. First, let's check average clustering:

    ```
    nx.average_clustering(G)
    0.007409464946430933
    ```

Clustering gives us a result of about `0.007`, which indicates that this is a sparse network. If clustering had returned a result of `1.000`, then that would indicate that every node is connected with every other node in the network. From an SNA context, that would mean that every person and organization in the network knows and interacts with each other. In K-pop, this is certainly not the case. Not all musicians know their fans. Not all fans are friends with their favorite idols.

2. What does `density` look like?

    ```
    from networkx.classes.function import density
    density(G)
    0.001830685967059492
    ```

Density gives us a result of about `0.002`, which further validates the sparsity of this network.

Let's not move on just yet. I want to make sure these concepts are understood. Let's construct a fully connected graph – a "complete" graph – with 20 nodes and repeat the steps from the preceding paragraphs. NetworkX has some handy functions for generating graphs, and we will use `nx.complete_graph` for this demonstration:

1. Let's build the graph!

    ```
    G_conn = nx.complete_graph(n=20)
    ```

2. First, let's investigate the size of the network:

    ```
    nx.info(G_conn)
    'Graph with 20 nodes and 190 edges'
    ```

Great. We have a network with 20 nodes, and those 20 nodes have 190 edges.

3. Is this actually a fully connected network, though? If it is, then we should get `1.0` for both clustering and density:

    ```
    nx.average_clustering(G_conn)
    1.0
    density(G_conn)
    1.0
    ```

4. Perfect. That's exactly what we expected. But what does this network look like? Let's use the same function we've been using throughout this book to draw the visualization:

    ```
    draw_graph(G_conn, edge_width=0.3)
    ```

This will draw our network without any node labels.

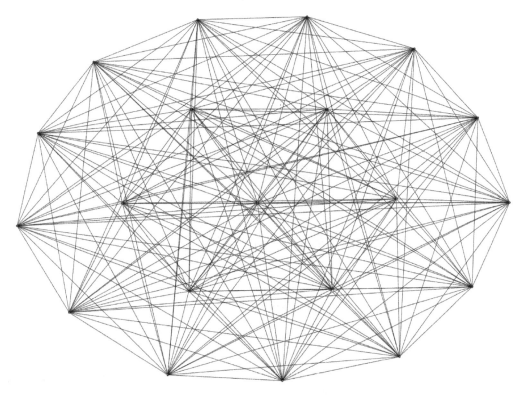

Figure 7.1 – Complete graph

As you can see in the network visualization, every node links with every other node. This is a fully connected network. Our K-pop network is a sparsely connected network, so the visualization will look very different.

Network visualization and thoughts

We know what a fully connected network looks like, and we know that the K-pop social network is sparsely connected, but what does that actually look like? Let's take a look:

```
draw_graph(G, node_size=1, show_names=False)
```

This will create a network visualization with nodes and edges, but without labels.

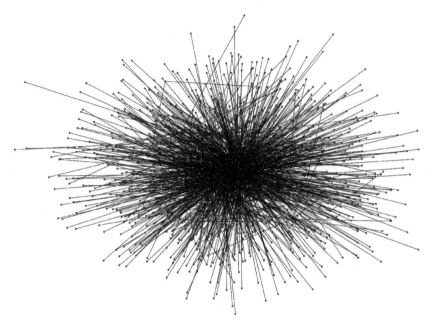

Figure 7.2 – K-pop network

One thing to notice is that even with only a thousand nodes, this still takes a few seconds to render, and it's impossible to pull any real insights out of the network. We can see a bunch of small dots, and we can see a bunch of lines from those small dots to other small dots. We can also notice that there is a core to the network and that the sparsity of the network increases the further we go toward the outskirts of the network. The idea of network layers will be explored later in this chapter. The point is that there's very little we can do with this visualization other than consider that it looks cool. At least we can visualize it, and in the later section of this chapter, I will explain how we can "peel the onion" to understand the various layers in the network.

But just to show something now, here is a very quick way to remove every node that only has a single edge, which is most of the network. If you do this, you can very quickly denoise a network. This is a huge time saver, as my previous approach for doing the exact same thing was to do the following:

1. Identify every node with a single edge, using a list comprehension.

2. Remove it from the network.

This one line of code removes the need for any of that. K_core converts the G graph into another graph that only contains nodes with two or more edges:

```
draw_graph(nx.k_core(G, 2), node_size=1, show_names=False)
```

Easy. How does the network look now?

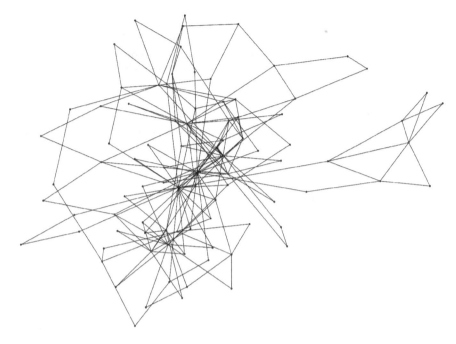

Figure 7.3 – K-pop network simplified

I hope that you can see that this one single step quickly brought out the structure of the network that exists underneath all of those nodes that only have a single edge. There are several ways to simplify a network, and I use this method frequently.

Important nodes

We now have an understanding of the general shape of the network, but we are interested in knowing who the most important people and organizations are. In network science, there is such a thing as **centrality scores** that give an indication of the importance of nodes in a network based on where they are placed and how information flows. NetworkX offers dozens of different centrality measures. You can learn about them at https://networkx.org/documentation/stable/reference/algorithms/centrality.html.

I will introduce a few of the centralities that I frequently use, but these are not necessarily the most important ones. Each centrality is useful for uncovering different contexts. The founders of Google also created their own centrality, famously known as **PageRank**. PageRank is a go-to centrality for many data professionals, but it may not be enough. To be thorough, you should understand the importance of nodes based both on how they are connected as well as how information moves. Let's explore a few different ways of gauging the importance of nodes in a network.

Degrees

The easiest way to judge the importance of somebody or something in a network is based on the number of connections between it and other nodes. Thinking about popular social networks such as Twitter or Facebook, influencers are often very well-connected, and we are suspicious of accounts that have very few connections. We are taking that concept and attempting to pull this insight from our network via code.

In a network, an entity (person, place, organization, and so on) in a network is called a node, and a relationship between one node and another is called an edge. We can count the number of edges each node has by investigating the degree counts of nodes in a network:

```
degrees = dict(nx.degree(G))
degrees
{'@kmg3445t': 1,
 '@code_kunst': 13,
 '@highgrnd': 1,
 '@youngjay_93': 1,
 '@sobeompark': 1,
 '@justhiseung': 1,
 '@hwajilla': 1,
 '@blobyblo': 4,
 '@minddonyy': 1,
 '@iuiive': 1,
 '@wgyenny': 1,
 ...
}
```

Now, we have a Python dictionary of nodes and their degree count. If we throw this dictionary into a pandas DataFrame, we can sort it and visualize the degree counts easily:

1. First, let's load it into a pandas DataFrame and sort by degrees in descending order (high to low):

    ```
    degree_df = pd.DataFrame(degrees, index=[0]).T
    degree_df.columns = ['degrees']
    degree_df.sort_values('degrees', inplace=True,
    ascending=False)
    degree_df.head()
    ```

This will show a DataFrame of Twitter accounts and their degrees.

	degrees
@b_hundred_hyun	128
@zanelowe	94
@haroobomkum	86
@spotifykr	80
@itzailee	79

Figure 7.4 – pandas DataFrame of node degrees

2. Now, let's create a horizontal bar chart for some quick insights:

```
import matplotlib.pyplot as plt
title = 'Top 20 Twitter Accounts by Degrees'
_ = degree_df[0:20].plot.barh(title=title, figsize=(12,7))
plt.gca().invert_yaxis()
```

This will visualize Twitter account connections by degrees.

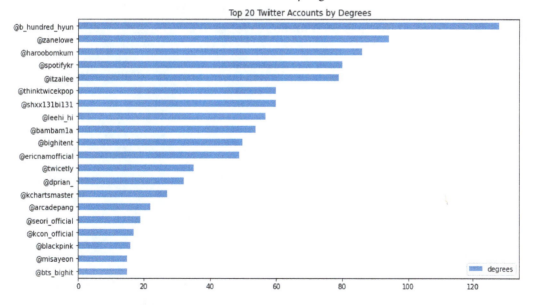

Figure 7.5 – Horizontal bar chart of Twitter accounts by degrees

One thing that stands out is that the number of degrees very quickly drops off, even when comparing the 20 most connected nodes. There is a significant drop-off even after the most connected node. The most connected node in the network belongs to singer/songwriter/actor Byun Baek-hyun – better known as Baekhyun – from the group Exo. That's interesting. Why is he so connected? Are people connecting to him, or does he connect to other people? Each insight tends to draw out more questions that can be explored. Write them down, prioritize by value, and then you can use those questions for deeper analysis.

Degree centrality

Degree centrality is similar to judging importance based on the number of degrees a node has. Degree centrality is the fraction of nodes in a network that a node is connected to. The more degrees a node has, the higher the fraction of nodes they will be connected to, so degrees and degree centrality can really be used interchangeably:

1. We can calculate the degree centrality of every node in the network:

```
degcent = nx.degree_centrality(G)
degcent
{'@kmg3445t': 0.0008605851979345956,
 '@code_kunst': 0.011187607573149742,
 '@highgrnd': 0.0008605851979345956,
 '@youngjay_93': 0.0008605851979345956,
 '@sobeompark': 0.0008605851979345956,
 '@justhiseung': 0.0008605851979345956,
 '@hwajilla': 0.0008605851979345956,
 '@blobyblo': 0.0034423407917383822,
 '@minddonyy': 0.0008605851979345956,
 '@iuiive': 0.0008605851979345956,

 ...

}
```

2. We can use this to create another pandas DataFrame, sorted by degree centrality in descending order:

```
degcent_df = pd.DataFrame(degcent, index=[0]).T
degcent_df.columns = ['degree_centrality']
degcent_df.sort_values('degree_centrality', inplace=True,
ascending=False)
degcent_df.head()
```

This will show a dataframe of Twitter accounts and their degree centralities.

	degree_centrality
@b_hundred_hyun	0.110155
@zanelowe	0.080895
@haroobomkum	0.074010
@spotifykr	0.068847
@itzailee	0.067986

Figure 7.6 – pandas DataFrame of nodes' degree centrality

3. Finally, we can visualize this as a horizontal bar chart:

```
title = 'Top 20 Twitter Accounts by Degree Centrality'
_= degcent_df[0:20].plot.barh(title=title,
figsize=(12,7))
plt.gca().invert_yaxis()
```

This will draw a horizontal bar chart of Twitter accounts by degree centrality.

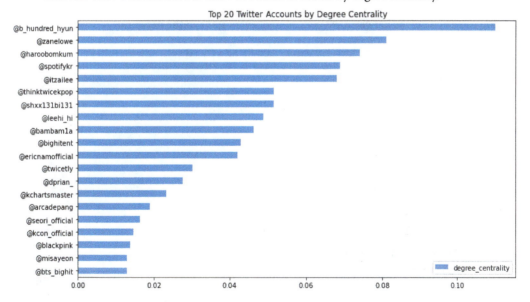

Figure 7.7 – Horizontal bar chart of Twitter accounts by degree centrality

Did you notice that the bar charts for degrees and degree centrality look identical other than the value? This is why I say that they can be used interchangeably. The use of degrees will likely be easier to explain and defend.

Betweenness centrality

Betweenness centrality has to do with how information flows through a network. If a node is positioned between two other nodes, then information from either of those two nodes must be passed through the node that sits between them. Information flows through the node that is sitting in the middle. That node can be seen as a bottleneck, or a place of advantage. It can give a strategic advantage to have the information that others need.

Usually, though, nodes with high betweenness centrality are situated between many nodes, not just two. This is often seen in a start network, where a core node is connected to dozens of other nodes or more. Consider an influencer on social media. That person may be connected to 22 million followers, but those followers likely do not know each other. They certainly know the influencer (or are an inauthentic bot). That influencer is a central node, and betweenness centrality will indicate that.

Before we see how to calculate betweenness centrality, please note that betweenness centrality is very time-consuming to calculate for large or dense networks. If your network is large or dense and is causing betweenness centrality to be so slow as to no longer be useful, consider using another centrality to calculate importance:

1. We can calculate the betweenness centrality of every node in the network:

```
betwcent = nx.betweenness_centrality(G)
betwcent
{'@kmg3445t': 0.0,
 '@code_kunst': 0.016037572215773392,
 '@highgrnd': 0.0,
 '@youngjay_93': 0.0,
 '@sobeompark': 0.0,
 '@justhiseung': 0.0,
 '@hwajilla': 0.0,
 '@blobyblo': 0.02836579219003866,
 '@minddonyy': 0.0,
 '@iuiive': 0.0,
 '@wgyenny': 0.0,
 '@wondergirls': 0.0013446180439736057,
 '@wg_lim': 0.002686271108798427,
```

```
    . . .
  }
```

2. We can use this to create another pandas DataFrame, sorted by betweenness centrality in descending order:

```
betwcent_df = pd.DataFrame(betwcent, index=[0]).T
betwcent_df.columns = ['betweenness_centrality']
betwcent_df.sort_values('betweenness_centrality',
inplace=True, ascending=False)
betwcent_df.head()
```

This will show a dataframe of Twitter accounts and their betweenness centralities.

	betweenness_centrality
@youtube	0.193090
@spotifykr	0.175619
@kchartsmaster	0.167481
@blackpink	0.125805
@haroobomkum	0.121789

Figure 7.8 – pandas DataFrame of nodes' betweenness centrality

3. Finally, we can visualize this as a horizontal bar chart:

```
title = 'Top 20 Twitter Accounts by Betweenness
Centrality'
_ = betwcent_df[0:20].plot.barh(title=title,
figsize=(12,7))
plt.gca().invert_yaxis()
```

This will draw a horizontal bar chart of Twitter accounts by betweenness centrality.

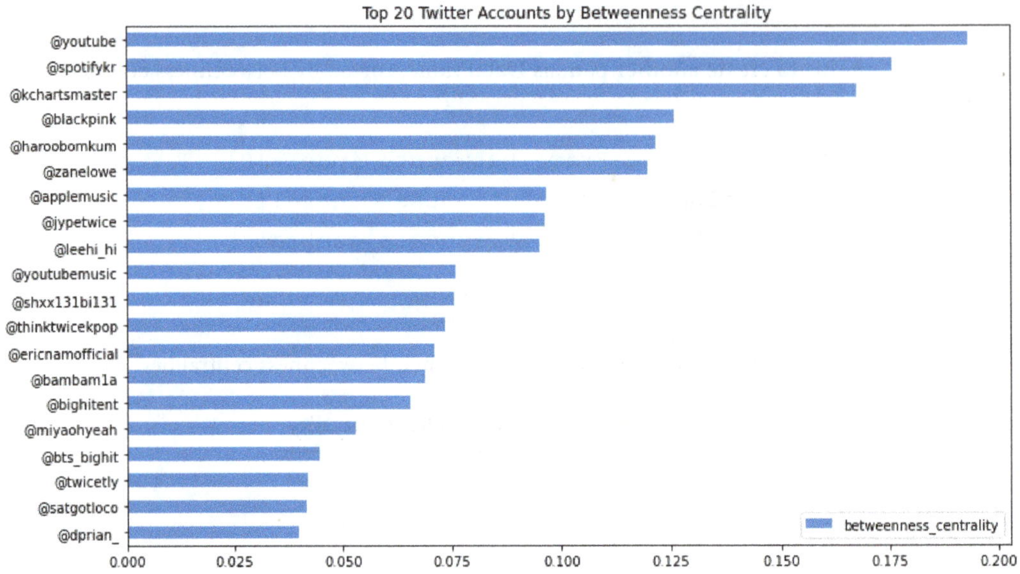

Figure 7.9 – Horizontal bar chart of Twitter accounts by betweenness centrality

Take note that the bar chart looks very different from the charts for degrees and degree centrality. Also note that @youtube, @spotifykr, and @kchartsmaster are the nodes with the highest betweenness centrality. This is likely because artists and others reference YouTube, Spotify, and KChartsMaster in their tweets. These nodes sit between nodes and other nodes.

Closeness centrality

Closeness centrality has to do with the closeness of nodes to other nodes, and that has to do with something known as the **shortest path**, which is computationally expensive (and slow) to compute for a large or dense network. As a result, closeness centrality may be even slower than betweenness centrality. If getting results from closeness centrality is too slow, due to the size and density of your own networks, you can choose another centrality for importance.

The shortest path will be explored in another chapter but has to do with the number of hops or handshakes it takes to get from one node to another node. This is a very slow operation, as there are many calculations at play:

1. We can calculate the closeness centrality of every node in the network:

    ```
    closecent = nx.closeness_centrality(G)
    closecent
    {'@kmg3445t': 0.12710883458078617,
    ```

```
'@code_kunst': 0.15176930794223495,
'@highgrnd': 0.12710883458078617,
'@youngjay_93': 0.12710883458078617,
'@sobeompark': 0.12710883458078617,
'@justhiseung': 0.12710883458078617,
'@hwajilla': 0.12710883458078617,
'@blobyblo': 0.18711010406907921,
'@minddonyy': 0.12710883458078617,
'@iuiive': 0.12710883458078617,
'@wgyenny': 0.07940034854856182,
...
}
```

2. We can use this to create another pandas DataFrame, sorted by closeness centrality in descending order:

```
closecent_df = pd.DataFrame(closecent, index=[0]).T
closecent_df.columns = ['closeness_centrality']
closecent_df.sort_values('closeness_centrality',
inplace=True, ascending=False)
closecent_df.head()
```

This will show a dataframe of Twitter accounts and their closeness centralities.

	closeness_centrality
@blackpink	0.247134
@youtube	0.238254
@kchartsmaster	0.230364
@spotifykr	0.229991
@leehi_hi	0.222560

Figure 7.10 – pandas DataFrame of nodes' closeness centrality

3. Finally, we can visualize this as a horizontal bar chart:

```
title = 'Top 20 Twitter Accounts by Closeness Centrality'
_ = closecent_df[0:20].plot.barh(title=title,
figsize=(12,7))
plt.gca().invert_yaxis()
```

This will draw a horizontal bar chart of Twitter accounts by closeness centrality.

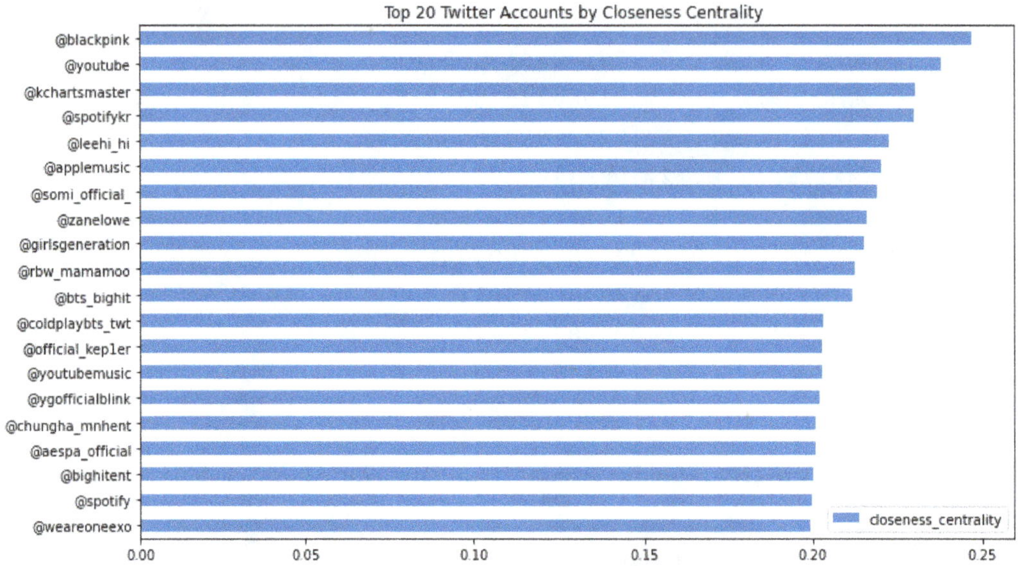

Figure 7.11 – Horizontal bar chart of Twitter accounts by closeness centrality

Take note that the results look different than every other centrality we have looked at. `@blackpink` is in the top spot, followed by `@youtube`, `@kchartsmaster`, and `@spotifykr`. BLACKPINK is a well-known K-pop group, and they are well-connected in the K-pop network, allowing them reach and influence. Other K-pop artists may want to investigate what it is that BLACKPINK is doing that puts them in a strategically advantageous network position.

PageRank

Finally, PageRank is the algorithm behind Google Search. The creators of Google wrote about it in 1999 in this paper: `http://ilpubs.stanford.edu:8090/422/1/1999-66.pdf`. If you have ever googled anything, the results that are returned to you are partially due to PageRank, though the search has likely evolved significantly since 1999.

The PageRank mathematical formula considers the number of inbound and outbound degrees of not only a node in question but of the linking nodes as well. It is because of this that **Search Engine Optimization (SEO)** has become a thing, as it became known that to get top Google positioning, a website should have as many inbound links as possible while also linking to other sources of information. For more information on the mathematics behind PageRank, check the PDF from Stanford University.

PageRank is a very fast algorithm, suitable for large and small networks, and very useful as an *importance* metric. Many graph solutions provide PageRank capabilities in their tools, and many people treat PageRank as their preferred centrality. Personally, I believe that you should know several centralities, where they are useful, and what their limitations are. PageRank is useful even for large and dense networks, so I recommend that it be included anytime you are doing any centrality analysis:

1. We can calculate the PageRank score of every node in the network:

```
pagerank = nx.pagerank(G)
pagerank
{'@kmg3445t': 0.00047123124840596525,
 '@code_kunst': 0.005226313735064201,
 '@highgrnd': 0.00047123124840596525,
 '@youngjay_93': 0.00047123124840596525,
 '@sobeompark': 0.00047123124840596525,
 '@justhiseung': 0.00047123124840596525,
 '@hwajilla': 0.00047123124840596525,
 '@blobyblo': 0.0014007295303692594,
 '@minddonyy': 0.00047123124840596525,
 ...
}
```

2. We can use this to create another pandas DataFrame, sorted by PageRank in descending order:

```
pagerank_df = pd.DataFrame(pagerank, index=[0]).T
pagerank_df.columns = ['pagerank']
pagerank_df.sort_values('pagerank', inplace=True,
ascending=False)
pagerank_df.head()
```

This will show a dataframe of Twitter accounts and their PageRank scores.

	pagerank
@b_hundred_hyun	0.050979
@zanelowe	0.036025
@haroobomkum	0.033742
@itzailee	0.031641
@spotifykr	0.026531

Figure 7.12 – pandas DataFrame of nodes' PageRank scores

3. Finally, we can visualize this as a horizontal bar chart:

```
title = 'Top 20 Twitter Accounts by Page Rank'
_= pagerank_df[0:20].plot.barh(title=title,
figsize=(12,7))
plt.gca().invert_yaxis()
```

This will draw a horizontal bar chart of Twitter accounts by PageRank.

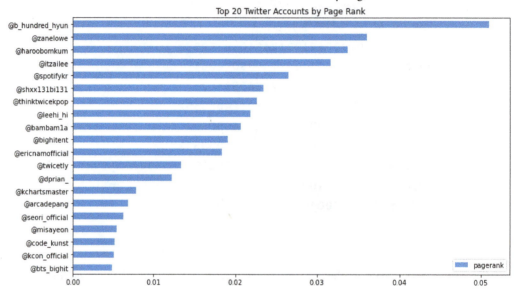

Figure 7.13 – Horizontal bar chart of Twitter accounts by page rank

These results actually look very similar to the bar chart from degrees and degree centrality. Once again, Baekhyun from Exo is in the top position.

Edge centralities

Before concluding this section on centralities, I want to point out that you are not limited to centralities for nodes. There are also centralities for edges. For instance, **edge betweenness centrality** can be used to identify the edge that sits between the most nodes. If you were to snip the edge that sits between most nodes, often the network would be split into two large pieces, called **connected components**. This can actually be useful for identifying communities or emerging trends. We will explore that more in a later chapter.

Comparing centralities

To get a feel for how the different centralities differ, or to use multiple different centralities together (for instance, if building an ML classifier and wanting to use graph metrics), it can be useful to combine the different centralities together into a single pandas DataFrame. You can easily do so with the pandas `concat` function:

```
combined_importance_df = pd.concat([degree_df, degcent_df,
betwcent_df, closecent_df, pagerank_df], axis=1)
combined_importance_df.head(10)
```

This will combine all of our centrality and PageRank DataFrames into one unified DataFrame. This will make it easier for us to compare different types of centralities.

	degrees	degree_centrality	betweenness_centrality	closeness_centrality	pagerank
@b_hundred_hyun	128	0.110155	0.012050	0.110155	0.050979
@zanelowe	94	0.080895	0.119831	0.216054	0.036025
@haroobomkum	86	0.074010	0.121789	0.197473	0.033742
@spotifykr	80	0.068847	0.175619	0.229991	0.026531
@itzailee	79	0.067986	0.004568	0.067986	0.031641
@thinktwicekpop	60	0.051635	0.073378	0.162288	0.022711
@shxx131bi131	60	0.051635	0.075669	0.165467	0.023469
@leehi_hi	57	0.049053	0.094873	0.222560	0.021838
@bambam1a	54	0.046472	0.068909	0.173223	0.020710
@bighitent	50	0.043029	0.065514	0.200317	0.019042

Figure 7.14 – pandas DataFrame of combined importance metrics

You may notice that if you rank by the different types of centralities, some have very similar results, and others are very different. I'll leave you with this: there is no single centrality to rule them all. They are different, and they should be used in different situations. If you are mapping out information flow, then betweenness centrality is very useful, so long as the network is of a manageable size. If you just want to see which nodes in a network are most connected, this is easiest to do by just investigating degrees. If you want to understand which nodes are situated closest to every other node, try closeness centrality. And if you want one algorithm that does a pretty good job at identifying important nodes and is performant even on large networks, try PageRank:

```
combined_importance_df.sort_values('pagerank', ascending=False)
[0:10]
```

This will show a DataFrame of Twitter accounts and combined network centralities and PageRank scores.

	degrees	degree_centrality	betweenness_centrality	closeness_centrality	pagerank
@b_hundred_hyun	128	0.110155	0.012050	0.110155	0.050979
@zanelowe	94	0.080895	0.119831	0.216054	0.036025
@haroobomkum	86	0.074010	0.121789	0.197473	0.033742
@ltzallee	79	0.067986	0.004568	0.067986	0.031641
@spotifykr	80	0.068847	0.175619	0.229991	0.026531
@shxx131bi131	60	0.051635	0.075669	0.165467	0.023469
@thinktwicekpop	60	0.051635	0.073378	0.162288	0.022711
@leehi_hi	57	0.049053	0.094873	0.222560	0.021838
@bambam1a	54	0.046472	0.068909	0.173223	0.020710
@bighitent	50	0.043029	0.065514	0.200317	0.019042

Figure 7.15 – pandas DataFrame of combined importance metrics sorted by PageRank

Just know that even PageRank and betweenness centrality can give very different results, so you should learn several different ways of determining importance and know what you are trying to do. These are very unfamiliar for beginners, but don't be afraid. Jump in and learn. The documentation and linked journals on NetworkX's documentation will be enough to help you get started.

Centralities are probably the most unusual section of this chapter if you are just getting started with social network analysis and network science. From this point on in the chapter, concepts should be less unusual.

Visualizing subgraphs

Often, in network analysis, we will want to see just a portion of the network, and how nodes in that portion link to each other. For instance, if I have a list of 100 web domains of interest or social media accounts, then it may be useful to create a subgraph of the whole graph for analysis and visualization.

For the analysis of a subgraph, everything in this chapter is still applicable. You can use centralities on subgraphs to identify important nodes in a community, for instance. You can also use community detection algorithms to identify communities that exist in a subgraph when the communities are unknown.

Visualizing subgraphs is also useful when you want to remove most of the noise in a network and investigate how certain nodes interact. Visualizing a subgraph is identical to how we visualize whole networks, ego graphs, and temporal graphs. But creating subgraphs takes a tiny bit of work. First, we need to identify the nodes of interest, then we need to construct a subgraph containing only those nodes, and finally, we will visualize the subgraph:

1. As an example, let's choose the 100 nodes from the network that have the highest PageRank scores:

    ```
    subgraph_nodes = pagerank_df[0:100].index.to_list()
    subgraph_nodes
    ['@b_hundred_hyun',
     '@zanelowe',
     '@haroobomkum',
     '@itzailee',
     '@spotifykr',
     '@shxx131bi131',
     '@thinktwicekpop',
     '@leehi_hi',
     '@bambam1a',
     '@bighitent',
     '@ericnamofficial',
     '@twicetly',
     ...
    ]
    ```

    That was easy. I'm only showing a few of the nodes, as this scrolls down the screen quite a way.

2. Next, I can construct a subgraph, like so:

    ```
    G_sub = G.subgraph(subgraph_nodes)
    ```

3. And finally, I can visualize it, the same way I would visualize any other network:

    ```
    draw_graph(G_sub, node_size=3)
    ```

In this example, I have left off the node names, but I could have just as easily added them. I felt it would make a cleaner visualization for this example.

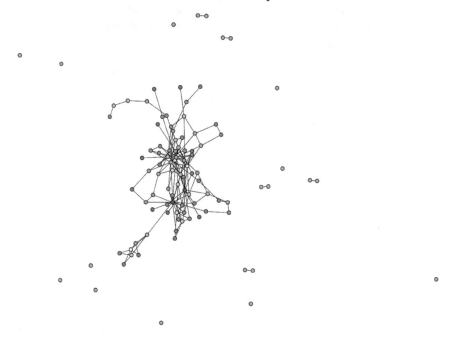

Figure 7.16 – Subgraph visualization of the top 100 K-pop Twitter accounts by PageRank

That's it. There's really not a lot to know about subgraph creation, other than that it is doable and how to do it. It is a simple process, once you know how.

Investigating islands and continents – connected components

If you take a look at the subgraph visualization, you may notice that there is one large cluster of nodes, a few small islands of nodes (two or more edges), and several isolates (nodes with no edges). This is common in many networks. Very often, there is one giant supercluster, several medium-sized islands, and so many isolates.

This presents challenges. When some people are new to network analysis, they will often visualize the network and use PageRank to identify important nodes. That is not nearly enough, for anything. There are so many different ways to cut the noise from networks so that you can extract insights, and I will show you several throughout the course of this book.

But one very simple way to cut through the noise is to identify the continents and islands that exist in a network, create subgraphs using them, and then analyze and visualize those subgraphs.

These *continents* and *islands* are formally called **connected components**. A connected component is a network structure where each node is connected to at least one other node. NetworkX actually allows for isolates to exist in their own connected components, which is strange to me, as isolates are not connected to anything other than possibly themselves (self-loops exist).

Finding all of the connected components that exist in a network is very easy:

```
components = list(nx.connected_components(G))
len(components)
```

I'm doing two things here: first, I load all connected components of our G graph into a Python list, and then I count the number of components that exist. There are 15 in the K-pop network.

Great, but which of these 15 are continents, and which are islands? Using a simple loop, we can count the number of nodes that exist in each connected component:

```
for i in range(len(components)):
    component_node_count = len(components[i])
    print('component {}: {}'.format(i, component_node_count))
```

This will give us a list of connected components and the number of nodes that are part of the connected component:

```
component 0: 909
component 1: 2
component 2: 3
component 3: 4
component 4: 2
component 5: 2
component 6: 80
component 7: 129
component 8: 3
component 9: 7
component 10: 4
component 11: 4
component 12: 2
component 13: 10
component 14: 2
```

Perfect. Notice that one of the components has 909 nodes. This is an example of one of those large continents that can exist in a network. Also, notice the components that have 80 and 129 nodes. This is significantly fewer than the number of nodes in the largest connected component, but it is still a significant number of nodes. I consider these as islands. Finally, notice that there are several other components that have between 2 and 10 nodes. These are like tiny islands.

Each of these connected components can be analyzed and visualized as a subgraph. For this exercise, to simplify visualization, I'll create a helper function to extend my main `draw_graph` function:

```
def draw_component(G, component, node_size=3, show_names=True)
    check_component = components[component]
    G_check = G.subgraph(check_component)
    return draw_graph(G_check, show_names=show_names, node_
size=node_size)
```

Let's try this out. Let's visualize a random component, component 13:

```
draw_component(G, component=13, node_size=5)
```

How does it render?

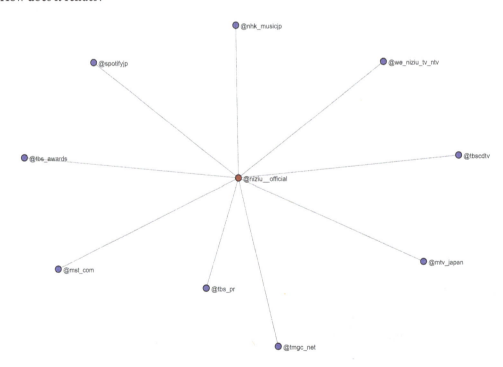

Figure 7.17 – Subgraph visualization of connected component #13

That looks good. We have successfully visualized a single component from the overall network. Let's visualize the largest component:

```
draw_component(G, 0, show_names=False, node_size=2)
```

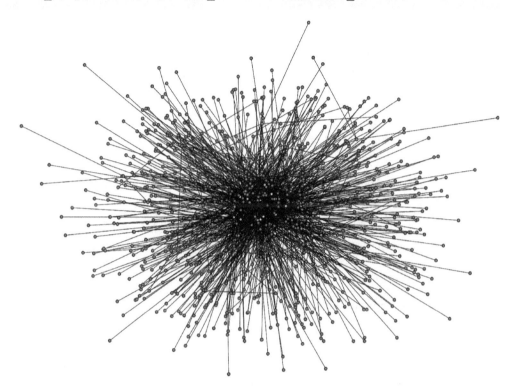

Figure 7.18 – Subgraph visualization of connected component #0

Once again, we are back to a giant, messy ball of yarn. We did successfully visualize it, though, and we can massively simplify it by removing all nodes with a single edge, for instance.

Connected components are a bit unusual, like centralities, but if you think of them as islands and continents that exist in a network, that really takes away a lot of the mystique. In summary, in a network, there are usually several connected components, and I consider them to be continents, islands, or isolates. In most networks, there is usually at least one large continent, several islands, and zero to many isolates. The number of isolates depends on how the graph was constructed. Using our NER approach from previous chapters, there are no isolates.

We will look at some more things we can do with connected components in *Chapter 9*.

Communities

Community detection algorithms are very useful in various forms of network analysis. In WNA, they can be used to identify communities that exist in the whole network. When applied to egocentric networks (ego graphs), they can reveal communities and cliques that exist around a single node, and in temporal networks, they can be used to watch communities evolve over time.

Community detection is common in SNA because communities of people exist in large populations, and it can be useful to identify those communities. There are various approaches to community detection, and there is quite a lot of information available on the internet for how community detection algorithms work. This book is about applied network science, so I am just going to demonstrate one, called the **Louvain algorithm**. Like centralities, there is no "best" algorithm. I have been in conversations where somebody pointed out a fringe algorithm that they were convinced was better, and I have been in conversations where people preferred Louvain.

You can learn more about the Louvain algorithm here: `https://python-louvain.readthedocs.io/en/latest/`.

1. The Louvain algorithm does not come with NetworkX. You will need to install it, which is as simple as the following:

    ```
    pip install python-louvain
    ```

2. After that, you can import the library for use with the following:

    ```
    import community as community_louvain
    ```

3. To save a lot of time and skip right past the math, the Louvain algorithm identifies various partitions (communities) that nodes are a part of. Visualizing these partitions is a bit trickier than our usual network visualizations, as `scikit-network` does not offer a lot of flexibility for coloring nodes. To save time, I'm going to return to my older network visualization practices and use NetworkX for visualization. Here is the code for drawing our graph and coloring the communities:

    ```
    def draw_partition(G, partition):

        import matplotlib.cm as cm
        import matplotlib.pyplot as plt

        # draw the graph
        plt.figure(3, figsize=(12,12))
        pos = nx.spring_layout(G)
    ```

```
    # color the nodes according to their partition
    cmap = cm.get_cmap('flag', max(partition.values()) +
1)

    nx.draw_networkx_nodes(G, pos, partition.keys(),
node_size=20, cmap=cmap, node_color=list(partition.
values()))

    nx.draw_networkx_edges(G, pos, alpha=0.5, width=0.3)


    return plt.show()
```

4. Now that we have the visualization function, we need to first identify partitions, and then we need to visualize the network. Let's do both of these together. I am using `resolution=2` after some tuning, as the community placement looks optimal:

```
partition = community_louvain.best_partition(G,
resolution=2)
draw_partition(G, partition)
```

How does it look?

Figure 7.19 – Visualization of community partitions

These images are messy but mesmerizing, to me. I can visually see easily distinguishable communities that I never noticed before. But what are they? What nodes are part of each community? It is simple to convert this partition list into a pandas DataFrame, and we can use that to identify communities, count the number of nodes that exist in each community, identify which community a node falls into, and visualize the individual communities:

1. First, let's create a pandas DataFrame from the partition list:

```
community_df = pd.DataFrame(partition, index=[0]).T
community_df.columns = ['community']
community_df.head()
```

How does it look now?

	community
@kmg3445t	0
@code_kunst	0
@highgrnd	0
@youngjay_93	0
@sobeompark	0

Figure 7.20 – pandas DataFrame of community partitions

2. This looks good. We can see that it is already sorted by partition number, which I am calling community. Now that this is in a pandas DataFrame, it is simple to count the number of nodes that belong to each community:

```
community_df['community'].value_counts()
```

This will give us a list of communities (the lefthand number) and a count of nodes that are part of the community (the righthand number):

```
21    170
10    133
14    129
16    104
2      91
3      85
13     80
23     70
0      66
```

15	55
4	51
22	48
1	36
17	10
19	7
9	4
20	4
5	4
8	3
18	3
12	2
11	2
7	2
6	2
24	2

We can easily see which communities have the most nodes. We should analyze and visualize these using subgraphs, as explained previously.

3. How do we identify the nodes that exist in each community, though? Let's just do this in pandas. Here is a simple helper function:

```
def get_community_nodes(commmunity_df, partition):

    community_nodes = community_df[community_
df['community']==partition].index.to_list()
    return community_nodes
```

4. We can use that function as it is, but I would prefer to take those community nodes, create a subgraph, and visualize it. Here is a helper function to do all of that:

```
def draw_community(G, community_df, partition, node_
size=3, show_names=False):
    community_nodes = get_community_nodes(community_df,
partition)
    G_community = G.subgraph(community_nodes)

    return draw_graph(G_community, node_size=node_size,
show_names=show_names)
```

Let's try one out:

```
draw_community(G, community_df, 1, show_names=True)
```

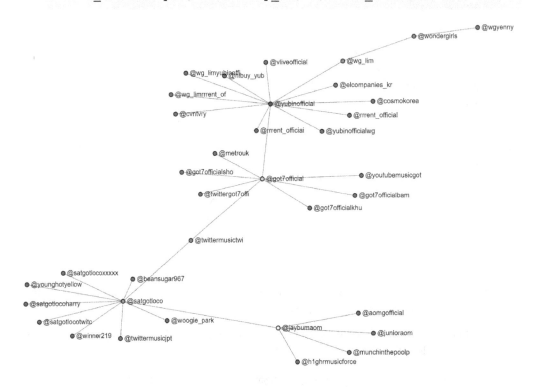

Figure 7.21 – Subgraph visualization of community #1

After running that, I can see this visualization. If you see something different, don't worry. When working with networks, things such as connected components and community numbers do not always end up in the same place on a subsequent run.

Very cool. This feels very similar to visualizing connected components, but communities are not necessarily islands or continents. Several communities can be found in large connected components, for instance. The algorithm looks for boundaries that separate groups of nodes, and then it labels communities accordingly.

If you work with networks, and especially if you are interested in identifying communities of people that exist in social networks, you will want to learn as much as you can about how to identify cliques and communities. Try different algorithms. I have chosen Louvain because it is fast and reliable, even on massive networks.

Bridges

In simple terms, bridges are nodes that sit between two different communities. These are typically easy to visually identify in small social networks, as there will be what looks like one or a few rubber bands or strengths that, if snipped, would allow the two groups to split apart. Just as a bridge allows people to traverse across the water from one piece of land to another, bridges in networks allow information to spread from one community to another. As a human, being a bridge is a powerful position to be in, as information and resources must flow through you to reach the other side.

In a complex network, bridges are more difficult to see visually, but they often exist, sitting between two communities. Our K-Pop network is pretty complex, so networks are less visible than they might be in a smaller social network, but they are there.

1. You can find bridges in a network like so:

    ```
    list(nx.bridges(G))
    [('@kmg3445t', '@code_kunst'),
     ('@code_kunst', '@highgrnd'),
     ('@code_kunst', '@youngjay_93'),
     ('@code_kunst', '@sobeompark'),
     ('@code_kunst', '@justhiseung'),
     ('@code_kunst', '@hwajilla'),
     ('@code_kunst', '@blobyblo'),
     ('@code_kunst', '@minddonyy'),
     ('@code_kunst', '@iuiive'),
     ('@code_kunst', '@eugenius887'),
     ...
    ]
    ```

2. This is a very long list of bridges, and I'm only showing a few of the rows, but we can use this along with pandas to identify the most important bridges:

    ```
    bridges = [s[0] for s in list(nx.bridges(G))]
    pd.Series(bridges).value_counts()[0:10]
    @b_hundred_hyun    127
    @zanelowe           90
    @haroobomkum        84
    @itzailee           78
    @spotifykr          60
    @shxx131bi131       57
    @thinktwicekpop     53
    ```

```
@leehi_hi          53
@bambam1a          49
@bighitent         46
```

3. One side effect of removing bridge nodes is that it can be similar to removing highly central notes – the network will shatter into a large group of isolates and a few smaller connected components. Let's take the 10 bridge nodes with the most edges and remove them:

```
cut_bridges = pd.Series(bridges).value_counts()[0:10].
index.to_list()
G_bridge_cut = G.copy()
G_bridge_cut.remove_nodes_from(cut_bridges)
```

4. After doing this, our network will likely look like a star has gone supernova, with debris flying out into space. Let's take a look:

```
draw_graph(G_bridge_cut, show_names=False)
```

This should draw a network without node labels.

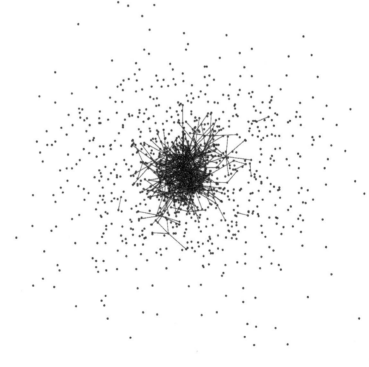

Figure 7.22 – Network visualization with top bridges cut

As we can see, there is still one densely connected component in the center of the network, a few tiny connected components consisting of a few nodes, and many individual isolates. Cutting bridges is not always so devastating. In other networks I have worked on, there's been a core nucleus consisting of two communities with a few nodes sitting between the two communities as bridges. When removing those bridges, the network core communities just split apart. There were few or no isolates.

There are reasons for identifying bridges. In a social network, these are nodes that information must flow through in order to reach communities on the other side. If you wanted to strategically place yourself in one of these networks, it would be wise to understand what the bridge is doing and then mimic what they have done, making connections with the people on each side. This can be a shortcut to power.

Similarly, if your goal was to disable a network, identifying and removing important bridges would cease the information flow from one community to another. It would be highly disruptive. This can be useful for disrupting dark networks (of crime, hate, and so on).

These are useful insights that can be extracted from networks, and they aren't as identifiable without network analysis. Identifying bridges and having a plan for what to do with them can provide a strategic advantage. You can use them to gain power, you can use them to disrupt, or you can use them to pull network communities apart for cleaner analysis.

Understanding layers with k_core and k_corona

Networks can be thought of as like onions, and they are often visualized similarly, with isolates drawn on the outside, nodes with a single edge rendered after that, then nodes with two edges, and on and on and on until the core of the network is reached. NetworkX allows two functions for peeling the onion, so to say: **k_core** and **k_corona**.

k_core

NetworkX's k_core function allows us to easily reduce a network to only nodes that have k or more edges, with "k" being a number between 0 and the maximum number of edges that any node has in a network. As a result, you get the "core" of a network that contains k or more edges. If you were to do k_core(G, 2), then this would return a graph containing nodes that have two or more edges, removing isolates and nodes with a single degree in one easy step.

That single step of denoising a network may not seem like a big deal, but doing this with list comprehensions or loops requires more steps, more thought, and more troubleshooting. This single step easily does the cleanup. As such, k_core(G, 2) is common in my code when I am most interested in the shape of the network that exists after removing isolates and single-edge nodes.

For instance, here is what our full K-pop network looks like when it is rendered. It is very difficult to see anything, as the single-edge nodes have turned the network visualization into a messy ball of yarn.

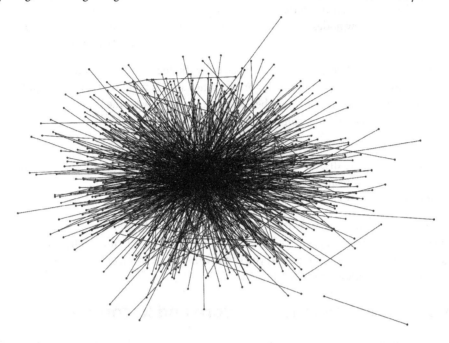

Figure 7.23 – Whole network visualization

However, we can easily remove all nodes that have fewer than two edges:

```
G_core = nx.k_core(G, 2)
```

How does the network look now?

```
draw_graph(G_core, show_names=True, node_size=3)
```

This should draw our `G_core` network with node labels.

Figure 7.24 – Whole network visualization with k_core and k=2

It should be obvious that this is much easier to interpret.

Learning about `k_core` was one of the most important moments for me as I learned how to analyze graphs and social networks. I used to denoise networks the less straightforward way, identifying nodes with fewer than two degrees, adding them to a list, and then removing them from a network. This single function has saved me so much time.

k_corona

Just as `k_core` allows us to extract the core of a network, `k_corona` allows us to investigate each layer of a network. `k_corona` is not about finding the core. It is about investigating what is happening in each layer of the network. For instance, if we only wanted to see nodes that have zero or one edges, we could do this:

```
G_corona = nx.k_corona(G, 1)
```

This would render as a bunch of isolates, and there will also likely be a few nodes that have one edge between them:

1. First, let's visualize the results of k_corona(G, 1):

    ```
    draw_graph(G_corona, show_names=False, node_size=2)
    ```

    This should render a network visualization of all nodes that have one or fewer edges. Nodes without any edges are called isolates and will appear as dots.

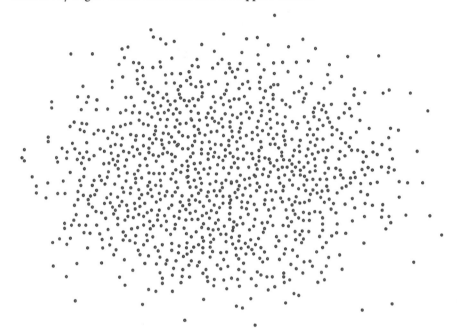

Figure 7.25 – Visualization of k_corona k=1 layer

As we can see, there are lots of isolates. Can you identify the few nodes that have a single edge between them? I can't. It's like reading that book, Where's Waldo? So, how do we identify the nodes in this layer that have an edge between them? How do we remove all of the nodes that have less than one edge? Think for a second.

2. That's right, we'll use k_core for the cleanup:

    ```
    G_corona = nx.k_corona(G, 1)
    G_corona = nx.k_core(G_corona, 1)
    ```

If we visualize this, we can see that there are five connected components, each containing two nodes, and each node has a single edge between it and another node that exists in that connected component.

```
draw_graph(G_corona, show_names=True, node_size=5, font_
size=12)
```

This will draw the G_corona network, with isolates removed, and with node labels showing.

○ @8_ohmygirl
○ @into__universe

○ @bol4_official
○ @shofarmusic

○ @day6official
○ @withdrama

○ @9muses_
○ @miiiliner_misog

○ @hunus_elris
○ @elris_official

Figure 7.26 – Visualization of k_corona k=1 layer, simplified

3. Is there an easy way to extract these nodes, so that we can use them for further analysis? Yes, easily:

```
corona_nodes = list(G_corona.nodes)
corona_nodes
```

This will show us a list of all of the nodes in corona_nodes:

```
['@day6official',
 '@9muses_',
 '@bol4_official',
 '@8_ohmygirl',
 '@withdrama',
 '@elris_official',
```

```
'@hunus_elris',
'@miiiiiner_misog',
'@into__universe',
'@shofarmusic']
```

4.   What does the second layer of the network look like, the layer where each node has two degrees? Are the nodes on this layer connected to each other? Let's create and render this visualization:

```
G_corona = nx.k_corona(G, 2)
draw_graph(G_corona, show_names=True, node_size=3)
```

This will render a network visualization of all nodes with two or fewer edges.

Figure 7.27 – Visualization of k_corona k=2 layer

It looks very similar to k_corona of layer one, but we can more easily see that a few nodes are connected to other nodes. We can also see that there are drastically fewer isolates in this layer. We could redo the k_core step for cleanup, but I think you get the point.

Personally, I don't use k_corona all that often. I have little interest in peeling networks, layer by layer, but the option is there, and maybe it can be more useful to you than it is to me. However, I use k_core practically every time I do anything with networks, for denoising a network, and for investigating the nucleus or nuclei that exist at the core of social networks. I recommend that you learn about both, but you may possibly have much more use for k_core than for k_corona. Still, k_corona opens up some interesting doors for analysis.

Challenge yourself!

Before concluding this chapter, I want to bring a challenge forward to you. You have learned how to create networks using text, what an edge list looks like and how to build one in pandas, how to create a network, and how to clean a network, and now you have an introduction to whole network analysis. You now have every tool required to start your journey into network analysis. I will explain how to do much more in later chapters, but you have all the tools you need to get started and get hooked on network analysis.

I want to challenge you to think about your own data, about the data you work with, about the social networks you play on, about the work networks you collaborate in, and more. I want you to consider how you could take these living networks, describe them in an edge list (it's just source and target columns), render a network visualization, and analyze the networks. You have the tools to do this, and networks are all around us. I recommend that when you are learning to investigate social networks and various networks, you use data that is actually interesting to you. You don't need to find a dataset online. You can easily create one yourself, or you can scrape social media, as we explored in earlier chapters.

I challenge you to stop at this chapter and play around for a while. Get lost in networks. Reread previous chapters in this book. Explore. Get weird. Have fun. This is my favorite way to learn. I take great enjoyment in building my own datasets and analyzing the things that I am most interested in.

Summary

We have gone a long way in this chapter. This chapter could be a book on its own, but my goal was to give a fast-paced tour of the things that you can do with networks. As I stated in the beginning, this will not be a math book. I want to unlock new capabilities and opportunities for you, and I feel that this chapter and this book can do this for you.

In this chapter, we covered a lot of ground: explaining whole network analysis, describing questions that can help with an analysis, and spending a lot of time actually doing network analysis. We looked at the network as a whole, but we also looked at node centralities, connected components, and layers.

In the next chapter, we are going to learn about egocentric network analysis. We call these **ego networks**, to be concise. In that chapter, we will zoom in on nodes of interest, to understand the communities and nodes that exist around them. You can think of egocentric network analysis as zooming in.

8
Egocentric Network Analysis

The previous chapter was a whirlwind. We covered so much material, learning how to visualize and analyze whole networks. In comparison, this chapter should feel much simpler. It will also be much shorter. In previous chapters, we learned how to get and create network data, how to build graphs from network data, how to clean graph data, and how to do interesting things such as identifying communities. In this chapter, we will be doing what is called **egocentric network analysis**.

The good news is that everything that was learned in the previous chapter applies to egocentric networks. Centralities can be useful for finding important nodes. Community algorithms can be useful for identifying communities. The great news is that there really isn't a lot that we need to cover in this chapter. Egocentric network analysis is simpler in scale as well as in scope. It's most important that I explain how to get started, show what you can do, and explain future steps you might want to take to go further in your analysis. As with any kind of analysis, there is always more that you can do, but we will keep things very simple in this chapter.

We will cover the following topics in the chapter:

- Doing egocentric network analysis
- Investigating ego nodes and connections
- Identifying other research opportunities

Technical requirements

In this chapter, we will mostly be using the Python libraries NetworkX and pandas. These libraries should be installed by now, so they should be ready for your use. If they are not installed, you can install Python libraries with the following:

```
pip install <library name>
```

For instance, to install NetworkX, you would do the following:

```
pip install networkx
```

In *Chapter 4*, we also introduced a `draw_graph()` function that uses both NetworkX and `scikit-network`. You will need that code any time that we do network visualization. You will need it for this chapter and most chapters in this book.

The code is available on GitHub: `https://github.com/PacktPublishing/Network-Science-with-Python`.

Egocentric network analysis

Egocentric network analysis is a form of network analysis that is useful for investigating the relationships that exist around a specific person in a social network. Rather than looking at the whole social network, we will zoom in on an individual and the individuals that this person interacts with. Egocentric network analysis uses a simpler form of a network called an **egocentric network** (**ego network**). From this point on, I will refer to these networks as *ego networks*.

In an ego network, there are two types of nodes: **ego** and **alters**. The ego node is the node of an individual that you are investigating. Alters, on the other hand, are all other nodes that exist in an ego network. If I were to make an ego network based on my own life, I would be the ego, and the people I know would be the alters. If I wanted to investigate the people who mention or are mentioned by the **@spotifykr** Twitter account in the K-pop social network from the previous chapter, I would create an ego network for spotifykr. spotifykr would be the ego, and all other nodes would be altered.

What's the point of this? Well, you can learn a lot about a person or organization based on the people who interact with that person. Like attracts like, in many cases. In my own life, most of my friends are engineers or data scientists, but my other friends are artists. I enjoy knowing creative and analytical people. Someone else might have a completely different makeup of the types of people they hang out with. Analyzing and visualizing ego networks can give us insights into relationships that we may be unable to see or notice living in the moment. We may have a hunch that certain types of relationships exist or how a person is being influenced, but being able to analyze and visualize an ego network itself is illuminating.

One nice thing about working with ego networks rather than whole networks is that ego networks tend to be much smaller and less complex than whole networks. That makes sense, as ego networks are subsets of a larger ecosystem. For instance, in a social network of millions of people, an ego network will focus on the ego node, and the alter nodes that surround it. As these networks are much smaller and less complex, this can allow for easy work with otherwise computationally expensive algorithms. There's less data to crunch. However, ego networks obviously scale based on the popularity of an individual. The ego network for a celebrity influencer will be much more complex than my own ego network. I don't have millions of followers.

Uses for egocentric network analysis

In social network analysis, ego networks are used to understand the relationships and communities that exist around a person. However, that is not at all the limit of what you can do with ego network analysis. I have used ego networks in a variety of different kinds of work to understand people's relationships, communication flow, and influence. I have also used networks to map out production data flows across a data center, and ego networks to investigate data flows and processes that exist around a piece of software or database table. If you can create a network, you can use ego networks to drill into the network for a closer look. You are not limited to analyzing people. You can use this to analyze families of malware, for instance. You can use this to understand how amplification works across social media. You can use this to inspect a component involved in a supply chain. You are limited by your own creativity and the data you are able to create or acquire. Anywhere that a network exists, ego networks can be used for analysis.

Explaining the analysis methodology

This chapter is going to be completely hands-on and repeatable. We will use a pre-built NetworkX network of the characters involved in the novel *Les Misérables*. I chose to use this network because it is large and complex enough to be interesting, but also because it has clear communities that can be seen inside various ego networks. This is an excellent network for practice with social network analysis.

The NetworkX network comes with weights, which allow us to draw thicker edges between nodes that have more interactions than those that have fewer interactions. However, for this analysis, I have chosen to drop the weights, as I want you to pay most attention to the alters and communities that exist around an ego node. For this analysis, I am more interested in the structure of the ego network and the communities that exist inside the ego network. I do recommend that you challenge yourself. As you make your way through the code of this chapter, maybe keep the weights rather than dropping them, and see how it affects the network visualizations.

We will start by doing a quick whole network spotcheck, just to see what the network looks like and to pick up centralities that could be useful for identifying interesting nodes that could use an ego network analysis.

After that, we'll look at four separate ego networks. We will start by learning just a little bit about the ego character from the novel, but we will not go into depth. Then, we will visualize the ego network both with and without its center. In an ego network, if you drop the ego node, it's called dropping the center. The ego node sits in the center. In an ego network, all alters have an edge between themselves and the ego node. If an alter is in an ego network, the alter has some form of relationship with the ego. So, what do you think will happen if you drop the ego node from an ego network? The ego network becomes simpler and may even break into pieces. This breakage is especially useful when it happens, as it becomes very easy to identify different communities, as they show as clusters of nodes. So, we will perform ego network analysis with the center dropped.

We will look to identify the alters that exist in an ego network, identify the most important alters, and we'll compare the density of each of the four ego networks. We will also look for communities as well as bridges that exist between communities.

Let's get started!

Whole network spot check

Before we can do anything with ego networks, we must first construct our graph. We have done this several times by now, so this should be familiar territory, but this time, we are going to use one of NetworkX's pre-built graphs. Loading a pre-built graph is simple:

```
import networkx as nx
G = nx.les_miserables_graph()
```

NetworkX has several other graphs available, so be sure to browse the documentation as you may find other networks of interest for your own work and learning.

This graph contains edge weights. While this is useful for understanding the number of interactions that take place between nodes, I have chosen to remove it from our graph so that we have clearer lines and so that we can focus on the ego networks themselves. These commands will convert the graph into a pandas edge list DataFrame – keeping only the source and target fields – and create a new graph using the DataFrame:

```
df = nx.to_pandas_edgelist(G)[['source', 'target']]
G = nx.from_pandas_edgelist(df)
```

Now that our revised graph is built, we can take a look at the number of nodes and edges:

```
nx.info(G)
'Graph with 77 nodes and 254 edges'
```

With only 77 nodes and 254 edges, this is a simple network, and we can easily visualize it in *Figure 8.1*:

```
draw_graph(G, font_size=12, show_names=True, node_size=4, edge_
width=1)
```

This produces the following network:

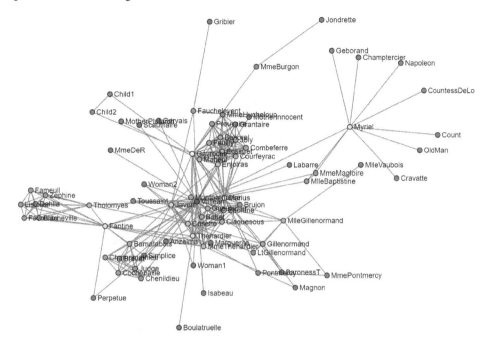

Figure 8.1 – Les Miserables whole network

This is good enough to move on, but I want to remind you of something we learned about in the previous chapter. We can use k_core to remove nodes from the visualization that have less than k nodes. In this case, I have chosen to not show nodes with fewer than two nodes:

```
draw_graph(nx.k_core(G, 2), font_size=12, show_names=True,
node_size=4, edge_width=1)
```

This will draw a network visualization, showing nodes that have two or more edges, effectively removing isolates and nodes with only a single edge. This will give us a quick understanding and preview of the structure of our network:

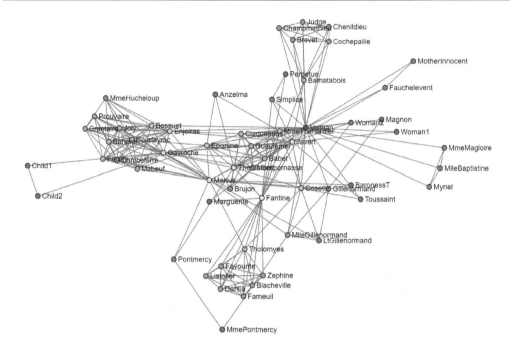

Figure 8.2 – Les Miserables whole network (k=2)

After visualizing a network, I tend to collect the PageRank scores of all nodes on a network. PageRank is a fast algorithm that scales well regardless of network size, so this is a good go-to algorithm for quickly identifying node importance, as discussed in the previous chapter. As a reminder, the `pagerank` algorithm calculates an importance score based on the number of incoming and outgoing edges that a node has. For this network, we are using an undirected graph, so `pagerank` is really calculating the score based on the number of edges that a node has, as there is no such thing as `in_degree` or `out_degree` in an undirected network. Here is how we can calculate `pagerank` and put the scores into a pandas DataFrame for quick analysis and visualization:

```
import pandas as pd
pagerank = nx.pagerank(G)
pagerank_df = pd.DataFrame(pagerank, index=[0]).T
pagerank_df.columns = ['pagerank']
pagerank_df.sort_values('pagerank', inplace=True,
ascending=False)
pagerank_df.head(20)
```

Let's visualize the `pagerank` algorithm calculation:

	pagerank
Valjean	0.075434
Myriel	0.042803
Gavroche	0.035764
Marius	0.030893
Javert	0.030303
Thenardier	0.027926
Fantine	0.027022
Enjolras	0.021880
Cosette	0.020611
MmeThenardier	0.019501
Bossuet	0.018957
Courfeyrac	0.018576
Eponine	0.017793
Mabeuf	0.017476
Joly	0.017197
Bahorel	0.017197
Gueulemer	0.016691
Babet	0.016691
Claquesous	0.016561
MlleGillenormand	0.016260

Figure 8.3 – Top 20 PageRank nodes in the Les Miserables network

A picture can help us see the difference more easily in the PageRank between each of the nodes:

```
pagerank_df.head(20).plot.barh(figsize=(12,8)).invert_yaxis()
```

The visualization is as follows:

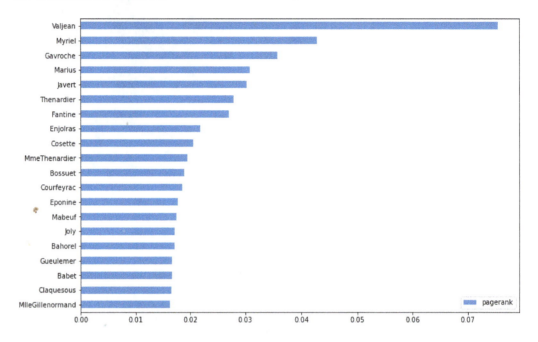

Figure 8.4 – Top 20 PageRank nodes visualized in the Les Miserables network

Excellent. We can clearly see that **Valjean** is a very important character in this story. We will definitely want to inspect Valjean's ego network, as well as the ego networks of **Myriel** and **Gavroche**. In an attempt to make sure that we don't get too similar ego networks, I have chosen **Joly** as the fourth character to inspect. Joly is much further down on the PageRank list.

This is all of the whole network analysis that we will be doing in this chapter. From this point on, we will be learning about ego networks. Let's go!

Investigating ego nodes and connections

In egocentric network analysis, we are interested in learning more about the communities that exist around a single node in a network. We are much less interested in the structure and makeup of the entire network. We are "zooming in," so to speak. We will use egocentric network analysis to inspect the communities that exist around core characters from *Les Miserables*.

Ego 1 – Valjean

According to Wikipedia, Jean Valjean is the protagonist of Les Miserables. Knowing this, it makes sense that Valjean has the highest PageRank score of all characters in the network. The main character of any story typically interacts with more characters than anyone else, and PageRank will reflect that. For the sake of this chapter, this is as much background digging as we will do per character. If you want to do a thorough analysis of the networks that exist in a piece of literature, you should go much deeper. In this chapter, I am most interested in showing how to work with ego networks.

Full ego network

Before we can analyze an ego network, we must first create one. In NetworkX, this is called `ego_graph` and one can be created by simply passing in the full graph as well as the name of the node that you would like to analyze:

```
ego_1 = nx.ego_graph(G, 'Valjean')
```

That's it. That's how to create an ego network. There are other parameters that you can pass in, but in practice, this is about as complicated as it gets:

```
draw_graph(ego_1, font_size=12, show_names=True, node_size=4,
edge_width=1)
```

We can now visualize the ego network:

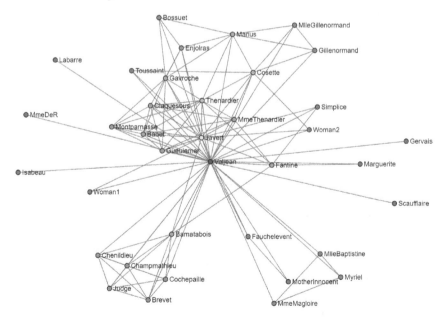

Figure 8.5 – The Valjean ego network

If you look closely, you should see that Valjean is in the center of his own ego network. This makes sense. The ego (Valjean) has some form of relationship with all of the alters (other nodes) that exist in the ego network.

Before moving on, there is something important that I want to point out. An ego network is just another network. Everything that we learned about in the previous chapter – *centralities*, *communities*, *degrees*, k_core, and k_corona – also applies to ego networks. Consider that the previous chapter mentioned that certain centralities are computationally expensive and time-consuming to run against whole networks. That is not always or even usually the case for ego networks, in my experience. Algorithms that were impractical for whole networks can be useful on ego networks, and can be applied with ease. There is an overlap between what can be useful for whole network analysis and egocentric network analysis. Everything that we learned in the previous chapter can be applied to ego networks.

Dropping the center from an ego network

One important option that is often overlooked is the ability to drop the center of an ego network. In simple terms, this means dropping Valjean out of his ego network. I find this very useful to do because when you drop a central node, often a network will shatter into pieces, making it much easier to identify the communities that exist in a network. We can drop the center out of an ego network like so:

```
ego_1 = nx.ego_graph(G, 'Valjean', center=False)
```

Now, that we have removed the center node – the ego – let's visualize the network again:

```
draw_graph(ego_1, font_size=12, show_names=True, node_size=4,
edge_width=1)
```

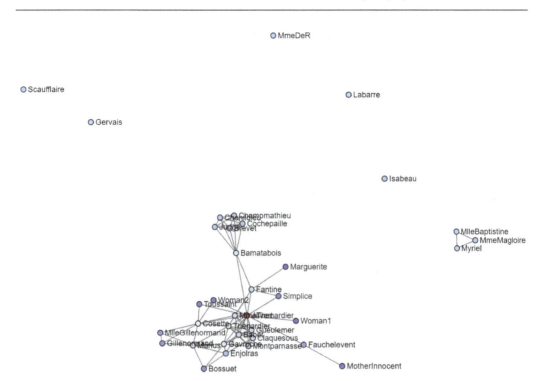

Figure 8.6 – The Valjean ego network with the center dropped

Compare this visualization to the previous one. What do you see? I can see that there is one large cluster of nodes that seems to have at least two communities as part of it at the top and to the left. I can also see a community of three nodes that has split off as an island on the right. And finally, I can see four isolate nodes. I hope it's clear to you that dropping the center can make it much easier to see these kinds of things.

Ego network (dropped center, denoised)

I often use k_core to denoise a network, and the same can be done for ego networks. Let's drop all nodes with less than one edge, effectively dropping the four isolate nodes:

```
draw_graph(nx.k_core(ego_1, 1), font_size=12, show_names=True,
node_size=4, edge_width=1)
```

Now, let's use the preceding code to visualize it:

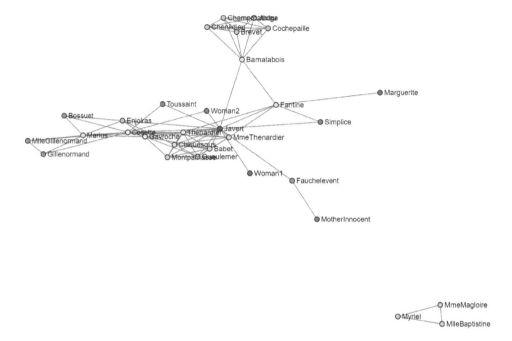

Figure 8.7 – The Valjean ego network with the center and isolates dropped

Now we have an even cleaner network, and it is clear that there are two clusters of nodes. A quick search on Wikipedia shows that **Myriel** is a bishop, **Magloire** is his servant and sister, and **Baptistine** is also his sister. It makes sense that they are part of their own community.

Alter list and amount

The same way we looked up nodes that exist in a whole network can also be used on an ego network. Rather than G.nodes, we'll use ego_1.nodes, as ego_1 is our ego network:

```
sorted(ego_1.nodes)
['Babet', 'Bamatabois', 'Bossuet', 'Brevet', 'Champmathieu',
'Chenildieu', 'Claquesous', 'Cochepaille', 'Cosette',
'Enjolras', 'Fantine', 'Fauchelevent', 'Gavroche', 'Gervais',
'Gillenormand', 'Gueulemer', 'Isabeau', 'Javert', 'Judge',
'Labarre', 'Marguerite', 'Marius', 'MlleBaptistine',
'MlleGillenormand', 'MmeDeR', 'MmeMagloire', 'MmeThenardier',
'Montparnasse', 'MotherInnocent', 'Myriel', 'Scaufflaire',
'Simplice', 'Thenardier', 'Toussaint', 'Woman1', 'Woman2']
```

There are two different ways that we can get the number of alters that exist in the ego network. Remember, we dropped the center node (the ego), so all remaining nodes are alters:

1. The first method is just to simply count the number of nodes in the network:

    ```
    len(ego_1.nodes)
    36
    ```

    That's fine, but what if we want to see the number of edges as well?

2. Let's just cut to the chase and use the `nx.info()` function instead of looking up how to get a list of all edges in a network, as this is easier:

    ```
    nx.info(ego_1)
    Graph with 36 nodes and 76 edges'
    ```

The whole network had 77 nodes, so clearly, the ego network is simpler.

Important alters

Getting a list of alters is one thing, but it is more useful if we can get a list of alters accompanied by some relevant centrality score to be able to gauge the importance of individual nodes in a network. Remember, there is no one centrality score to rule them all. We could have used PageRank, closeness centrality, betweenness centrality, or any of the other measures. These are the nodes that are connected to most other nodes in the ego network:

```
degcent = nx.degree_centrality(ego_1)
degcent_df = pd.DataFrame(degcent, index=[0]).T
degcent_df.columns = ['degree_centrality']
degcent_df.sort_values('degree_centrality', inplace=True,
ascending=False)
degcent_df.head(10)
```

Let's visualize this and take a look at our centralities:

	degree_centrality
Javert	0.457143
Thenardier	0.285714
Cosette	0.228571
Gavroche	0.228571
Babet	0.200000
Marius	0.200000
Gueulemer	0.200000
Bamatabois	0.200000
MmeThenardier	0.200000
Claquesous	0.200000

Figure 8.8 – The Valjean ego network alters' degree centrality

Javert has a clear lead over the other individuals with a `degree_centrality` of around `0.457`. That makes sense, considering how central Javert is in the ego network.

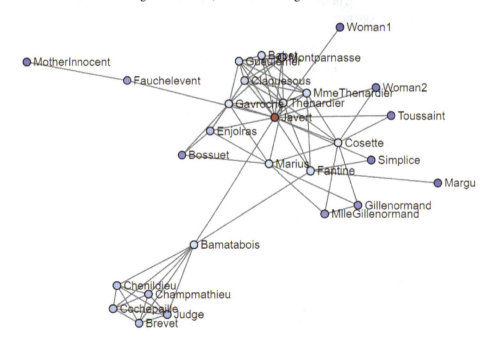

Figure 8.9 – Javert's network position in the Valjean ego network

The other alters with the highest centralities are much more difficult to see. **MmeThenardier** is in the center, poking out on the right.

Ego network density

Let's wrap up this ego network by getting the network density. Density has to do with the level of connectedness between all the nodes in a network. In order to have a density of 1.0, every single node would have an edge with every single other node in a network. To have a density of 0.0, a network would be made entirely of isolates, with no edges between any nodes. Can you guess roughly what the density would be for this network? It looks loosely connected to me, with a few pockets of densely connected communities. So, my guess would be a pretty low score. Let's use NetworkX to calculate the density:

```
nx.density(ego_1)
0.12063492063492064
```

A density of around 0.12 is a loosely connected network. I've calculated density because I want to use this to be able to compare the density of each of the ego networks.

> **Note**
>
> You might be wondering why we are looking at betweenness centrality and density or wondering what the relationship is between centrality and density. Centrality scores are useful for understanding the importance of a node in a network. Density tells us about the overall composition of the network. If a network is dense, then nodes are more connected than in a sparse network. Centrality and density scores are a way to get a quick understanding of a network.

Ego 2 – Marius

Wikipedia lists Marius Pontmercy as another protagonist in the novel. Let's look at his ego network next.

Full ego network

First, we'll construct the whole ego network without dropping the center:

```
ego_2 = nx.ego_graph(G, 'Marius')
draw_graph(ego_2, font_size=12, show_names=True, node_size=4,
edge_width=1)
```

Next, we'll visualize the whole ego network:

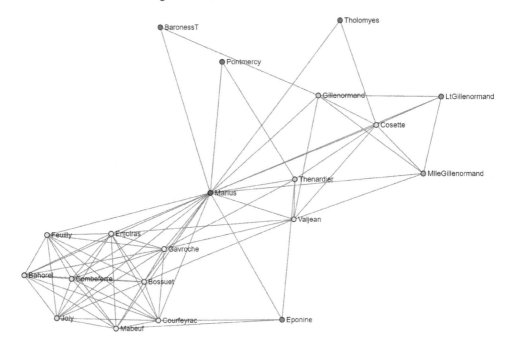

Figure 8.10 – Marius' ego network

Perfect. One thing is clear: this ego network looks completely different than Valjean's. Looking at the bottom left, I can see a densely connected community of individuals. Looking to the right, I can see a character that **Marius** has fond feelings for, but no spoilers.

Ego network (dropped center)

Let's drop the center (ego) out of this network and see how it looks:

```
ego_2 = nx.ego_graph(G, 'Marius', center=False)
draw_graph(ego_2, font_size=12, show_names=True, node_size=4,
edge_width=1)
```

This will draw our ego network:

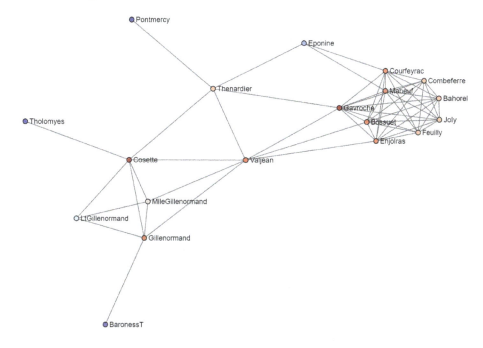

Figure 8.11 – Marius' ego network with the center dropped

Dropping the center also has drastically different results than when we did it for Valjean's ego network. In Valjean's case, four isolates detached from the network. In Marius' ego network, even with the center dropped, there are no isolates. The members of his ego network are connected well enough that removing Marius' node did not break the network apart. This network is resilient.

The densely connected community that exists in this ego network is also easily visible on the right. I can also see Valjean near the center of the network.

Previously, we used k_core to remove the isolate nodes so that we could look at the remaining nodes more easily after dropping the center. In Marius' ego network, we will skip that step. There are no isolates to remove.

Alter list and amount

Let's take a look at the alters that exist in Marius' ego network:

```
sorted(ego_2.nodes)
['Bahorel', 'BaronessT', 'Bossuet', 'Combeferre', 'Cosette',
 'Courfeyrac', 'Enjolras', 'Eponine', 'Feuilly', 'Gavroche',
 'Gillenormand', 'Joly', 'LtGillenormand', 'Mabeuf',
```

```
'MlleGillenormand', 'Pontmercy', 'Thenardier', 'Tholomyes',
'Valjean']
```

Next, let's get the number of nodes and edges the easy way:

```
nx.info(ego_2)
'Graph with 19 nodes and 57 edges'
```

Perfect. This is a very simple network.

Important alters

Now, let's see which alters sit in central positions. They are in powerful positions in the network:

```
degcent = nx.degree_centrality(ego_2)
degcent_df = pd.DataFrame(degcent, index=[0]).T
degcent_df.columns = ['degree_centrality']
degcent_df.sort_values('degree_centrality', inplace=True,
ascending=False)
degcent_df.head(10)
```

This will give us our degree centralities. Let's take a closer look!

	degree_centrality
Gavroche	0.555556
Mabeuf	0.500000
Bossuet	0.500000
Courfeyrac	0.500000
Enjolras	0.500000
Joly	0.444444
Combeferre	0.444444
Feuilly	0.444444
Bahorel	0.444444
Valjean	0.388889

Figure 8.12 – Marius' ego network alters' degree centrality

Wow, interesting. I would have suspected that Valjean would have been one of the most central nodes, but several people are ahead of him. Can you guess why? They are part of the densely connected community, and each of them is connected to more members of the ego network than **Valjean**. There should be a lot of information sharing in this community. Looking at Valjean's placement now, I can see that he is a central figure, but he is less connected than the members of the densely connected community.

Do note that several nodes have the same centrality score. That can happen. Centrality scores are just the result of math, not magic.

Ego network density

Finally, so that we can compare ego networks, let's calculate the density score:

```
nx.density(ego_2)
0.3333333333333333
```

Remember, Valjean's ego network had a density of around 0.12. Marius' ego network is almost three times as dense. This would explain why it did not shatter into pieces when Marius' central node was removed. A densely connected network is more resilient to breakage when central nodes are removed. This is important to remember when considering how real-world networks can be bolstered to provide greater availability. From a human standpoint, this community will continue to exist even when key nodes are removed.

Ego 3 – Gavroche

In *Les Miserables*, Gavroche is a young boy who lives on the streets of Paris. Keeping that in mind, I would imagine that his ego network would look very different from an adult or someone who is much more connected in society. Let's take a look.

Full ego network

First, let's visualize the network with the center intact:

```
ego_3 = nx.ego_graph(G, 'Gavroche')
draw_graph(ego_3, font_size=12, show_names=True, node_size=4,
edge_width=1)
```

This will render Gavroche's ego network:

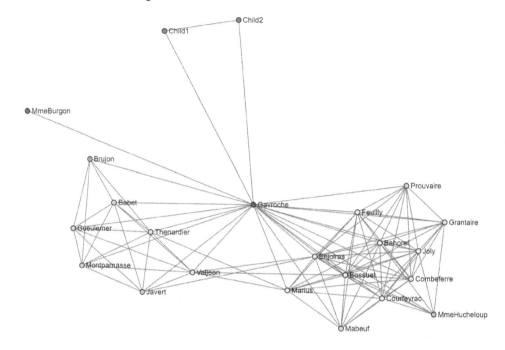

Figure 8.13 – Gavroche's ego network

Interesting. Gavroche is well connected. Knowing that he is a child who lives on the streets, it's interesting to see **Child1** and **Child2** listed at the top. The relationship between these three characters seems like it might be pretty interesting to understand. I can also see one person (**MmeBurgon**) whose node will become an isolate when Gavroche's center node is dropped. Finally, I see what looks like two communities on the bottom left and right of the ego network. These should become more clear after dropping the center of this network.

Ego network (dropped center)

Let's drop the center node and visualize the network again:

```
ego_3 = nx.ego_graph(G, 'Gavroche', center=False)
draw_graph(ego_3, font_size=12, show_names=True, node_size=4,
edge_width=1)
```

This will render Gavroche's ego network with the center dropped. This should make the separate communities more easily identifiable:

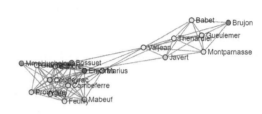

Figure 8.14 – Gavroche's ego network with the center dropped

Perfect. As expected, one node became an isolate, the two children formed their own little cluster, and the remaining connected component contains two separate communities of people. It could be interesting to use a community detection algorithm on the largest cluster, but let's move on. It's clear, though, that Valjean is positioned between these two communities.

There's not much point in removing isolates from this network, as there is only one, so let's push forward.

Alter list and amount

Let's see which other characters are a part of Gavroche's ego network. In other words, I want to know who Gavroche knows. This will tell us which characters play a part in his life:

1. Enter the following code:

    ```
    sorted(ego_3.nodes)
    ```

    This simple line will give us all of the nodes in Gavroche's ego network, sorted alphabetically:

    ```
    ['Babet', 'Bahorel', 'Bossuet', 'Brujon', 'Child1',
    'Child2', 'Combeferre', 'Courfeyrac', 'Enjolras',
    'Feuilly', 'Grantaire', 'Gueulemer', 'Javert', 'Joly',
    'Mabeuf', 'Marius', 'MmeBurgon', 'MmeHucheloup',
    'Montparnasse', 'Prouvaire', 'Thenardier', 'Valjean']
    ```

Great. I can see some familiar names, and I can clearly see `Child1` and `Child2` as well.

2. Next, let's see how many nodes and edges exist in this network, doing this the easy way:

```
nx.info(ego_3)
```

This should give us the following output:

```
'Graph with 22 nodes and 82 edges'
```

Wow. This is much more connected than the protagonist's own ego network. Valjean's ego network had 36 nodes and 76 edges. How do you think having fewer nodes and more edges will affect this network's density score?

Before we get to that, let's look at centralities.

Important alters

Once again, let's use degree centrality to see who the most connected people are in this ego network. These steps should start to look similar by now as they are. We are using different centralities to understand which nodes are important in our network:

```
degcent = nx.degree_centrality(ego_3)
degcent_df = pd.DataFrame(degcent, index=[0]).T
degcent_df.columns = ['degree_centrality']
degcent_df.sort_values('degree_centrality', inplace=True,
ascending=False)
degcent_df.head(10)
```

This will give us a DataFrame of characters by degree centrality. Let's look closer:

	degree_centrality
Enjolras	0.619048
Bossuet	0.571429
Joly	0.523810
Courfeyrac	0.523810
Bahorel	0.523810
Feuilly	0.476190
Marius	0.476190
Combeferre	0.476190
Grantaire	0.428571
Mabeuf	0.380952

Figure 8.15 – Gavroche's ego network alters' degree centrality

Wow, **Enjoiras** is one highly connected individual, and there are several other highly connected individuals as well. We can see them in the especially well-connected community.

Ego network density

Finally, let's calculate density so that we can compare ego networks:

```
nx.density(ego_3)
0.354978354978355
```

That is very interesting. Gavroche is a kid who lives on the streets of Paris, but his social network is denser than any of the others that we have seen before. Valjean's ego network density was around 0.12, Marius' was around 0.33, and Gavroche's is even higher. I never would have expected this. If I read this story, I am going to pay close attention to this character. He seems well connected, and I wonder how that plays out in the story.

Ego 4 – Joly

Near the beginning of this chapter, I chose four individuals for creating ego networks. The first three had high PageRank scores, and I specifically chose someone with a lower PageRank score for the fourth, as I was hoping for a drastically different ego network than the other three.

Joly is a medicine student, and I wonder if this will affect his ego network. Students often socialize with other students, so I'll investigate a few of his direct connections.

Full ego network

First, let's create and visualize an ego network with the center intact:

```
ego_4 = nx.ego_graph(G, 'Joly')
draw_graph(ego_4, font_size=12, show_names=True, node_size=4,
edge_width=1)
```

This will render Jolys ego network.

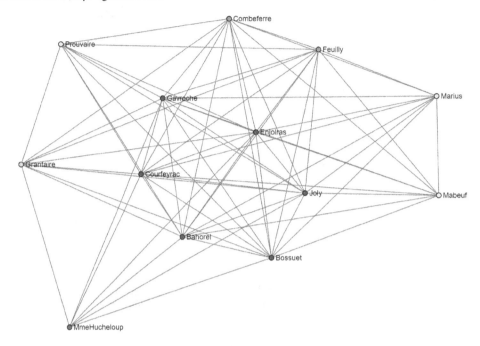

Figure 8.16 – Joly's ego network

Wow. Very often, when visualizing networks, one will just jump out at me as being really unique. Of the four that we have done for this chapter, this one stands out. It does not look like a typical ego network. It is densely connected. Everybody in Joly's ego network is well connected, and if we were to remove Joly's node, the ego network would likely change very little. Before we do this, let's take a look at a few of these individuals to see whether any of the others are either medicine students or members of Les Amis de l'ABC, which is an association of revolutionary students that Joly is a part of.

Bossuet

Bossuet Lesgle is known as the unluckiest student and is a member of Les Amis de l'ABC. Being both a student and a member of the revolutionary students, it makes sense that he is connected to Joly.

Enjoiras

Enjoiras is the leader of Les Amis de l'ABC. This connection also makes sense.

Bahoret

Bahoret is another member of Les Amis de l'ABC.

Gavroche

Gavroche does not appear to be a member of Les Amis de l'ABC, but he assists in fighting alongside them.

Even without knowing much of the story or even much about the individual characters, we can easily identify members of the same community by inspecting whole and ego networks as well as communities.

Ego network (dropped center)

Now, let's drop the center out of the ego network and see whether my hypothesis is correct that the ego network will not change very much because it is densely connected:

```
ego_4 = nx.ego_graph(G, 'Joly', center=False)
draw_graph(ego_4, font_size=12, show_names=True, node_size=4,
edge_width=1)
```

This will draw Joly's ego network with the center dropped. If there are separate communities in this network, they will be shown separately as clusters. If there is only one community, then that community will show as a single cluster:

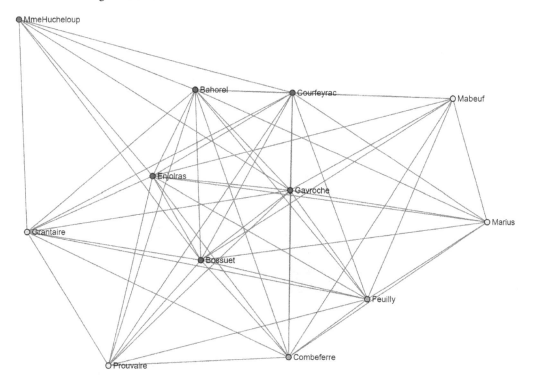

Figure 8.17 – Joly's ego network with the center dropped

That's very interesting. With Joly removed, the ego network is definitely intact, and central individuals remain central. This is a resilient network, and as they are a group of revolutionaries, it would be in their best interests to be resilient for their revolution to stand a chance.

There are no isolates, so there is no point in denoising this network.

Alter list and amount

Let's get a list of all of the individuals who are a part of this ego network:

```
sorted(ego_4.nodes)
```

This will give us a sorted list of characters who are part of this ego network:

```
['Bahorel', 'Bossuet', 'Combeferre', 'Courfeyrac', 'Enjolras',
 'Feuilly', 'Gavroche', 'Grantaire', 'Mabeuf', 'Marius',
 'MmeHucheloup', 'Prouvaire']
```

Compared to the other ego networks, there are very few individuals.

Let's get the number of nodes and edges:

```
nx.info(ego_4)
```

This should tell us a bit about the size and complexity of this ego network:

```
'Graph with 12 nodes and 57 edges'
```

This network has fewer nodes and fewer edges than other networks we've looked at, but it is visibly denser than any of the other networks.

Important alters

Let's take a look at the most important alters in the ego network:

```
degcent = nx.degree_centrality(ego_4)
degcent_df = pd.DataFrame(degcent, index=[0]).T
degcent_df.columns = ['degree_centrality']
degcent_df.sort_values( 'degree_centrality', inplace=True,
ascending=False)
degcent_df.head(10)
```

	degree_centrality
Bahorel	1.000000
Gavroche	1.000000
Bossuet	1.000000
Courfeyrac	1.000000
Enjolras	1.000000
Feuilly	0.909091
Combeferre	0.909091
Grantaire	0.818182
Mabeuf	0.727273
Prouvaire	0.727273

Figure 8.18 – Joly's ego network alters' degree centrality

These centralities are incredible, compared to the others that we have seen. The centrality scores of 1.0 for **Bahorel**, **Gavroche**, **Bossuet**, **Courfeyrac**, and **Enjoiras** means that they are connected to every single other person in this ego network. Even the individuals who do not have a centrality of 1.0 have very high-degree centralities. This is a well-connected network. How do you think that will affect the density score?

Ego network density

Let's calculate the density of this ego network. I suspect that this ego network will have a very high density:

```
nx.density(ego_4)
0.8636363636363636
```

That is incredibly high compared to the other density scores:

- Valjean's ego network had a density of around 0.12
- Marius's ego network was around 0.33
- Gavroche's ego network was around 0.35
- Joly's ego network was around 0.86

That is dramatically higher than any of the other density scores.

But this makes sense. Joly is both a student and a member of a revolutionary group. I would expect both of these groups to be well connected with other students and revolutionaries.

Insights between egos

We've looked at four different ego networks. We could have created an ego network for every single node in the network, but that would have been very time-consuming. When I do egocentric network analysis, I typically start by shortlisting several nodes of interest that I would like to investigate. I am typically looking for something, such as the following:

- Who is most connected
- Who has the most `out_degrees`
- Who has the highest `pagerank`
- Who has a tie to known opposition

In this chapter, we investigated characters from the novel *Les Miserables*, so I purposefully chose individuals with the highest PageRank scores, as I anticipated that they'd have interesting ego networks. This strategy worked very well.

There are a few takeaways from all of this that I'd like to leave you with before concluding this chapter:

- First, it is possible to bolster a network by adding connections, and this will make it resilient to failure. If one node is removed, the network can remain intact and not shatter into pieces. This has implications for many things. For the flow of information to remain constant, an information-sharing network will want to be resilient to attack, for instance. Where else might having a network be resilient to failure be of value?
- Second, removing the center node of an ego network can tell you a lot about the communities that exist in a network, as well as the resiliency of that network. When removing the center, which networks had isolates and islands? Which networks stayed completely intact? Which communities were we able to see?

Let's see what's next.

Identifying other research opportunities

When doing any kind of network analysis, it is important to know that there is always more that you can do; for instance, we could do the following:

- We could embed additional information into the graph, such as weights or node types (teacher, student, revolutionary, and so on)
- We could color nodes by node type for easier identification of communities
- We could size nodes by the number of degrees or the centrality score to make it easier to identify important nodes
- We could use directed networks to understand the directionality of information sharing

There is always more that you can do, but it is important to know that enough is enough. You should use what you need. For this chapter, I originally got stuck trying to do too much, and I lost time trying to figure out how to do things that weren't actually all that important to teaching this topic. Keep it simple and add only what you need and what is helpful. Add more as time permits if it is useful.

Summary

In this chapter, you learned a new kind of network analysis, called egocentric network analysis. I tend to call egocentric networks *ego networks*, to be concise. We've learned that we don't have to analyze a network as a whole. We can analyze it in parts, allowing us to investigate a node's placement in the context of its relationship with another node.

Personally, egocentric network analysis is my favorite form of network analysis because I enjoy investigating the level of the individual things that exist in a network. Whole network analysis is useful as a broad map, but with egocentric network analysis, you can gain a really intimate understanding of the various relationships between things that exist in a network. I hope you enjoyed reading and learning from this chapter as much as I enjoyed writing it. I hope this inspires you to learn more.

In the next chapter, we will dive into community detection algorithms!

9
Community Detection

In the last two chapters, we covered whole network analysis and egocentric network analysis. The former was useful for understanding the complete makeup of a complex network. The latter was useful for investigating the people and relationships that exist around an "ego" node. However, there's a missing layer that we have not yet discussed. Between whole networks and egos, communities exist. We are people, and we are part of a global population of humans that exist on this planet, but we are each also part of individual communities. For instance, we work in companies and as part of individual teams. Many of us have social interests, and we know people from participating in activities. There are layers to life, and we can use algorithms to identify the various communities that exist in a network, automatically.

This chapter contains the following sections:

- Introducing community detection
- Getting started with community detection
- Exploring connected components
- Using the Louvain method
- Using label propagation
- Using the Girvan-Newman algorithm
- Other approaches to community detection

Technical requirements

In this chapter, we will mostly be using the NetworkX and pandas Python libraries. These libraries should be installed by now, so they should be ready for your use. If they are not installed, you can install the Python libraries with the following:

```
pip install <library name>
```

For instance, to install NetworkX, you can use this:

```
pip install networkx
```

In *Chapter 4*, we also introduced a `draw_graph()` function that uses both NetworkX and Scikit-Network. You will need that code any time that we do network visualization. You will need it for this chapter, and most of the chapters in this book.

For community detection, we will also be using `python-louvain`. You can install it with the following:

```
pip install python-louvain
```

You can import it like this, which you will see later in this chapter:

```
from community import community_louvain
```

It is understandable if you're confused by the installation and import commands for `python-louvain`. The library name does not match the import library name. It's a useful library for community detection, so let's accept this as an oddity and move on.

Introducing community detection

Community detection is about identifying the various communities or groups that exist in a network. This is useful in social network analysis, as humans interact with others as part of our various communities, but these approaches are not limited to studying humans.

We can also use these approaches to investigate any kinds of nodes that interact closely with other nodes, whether those nodes are animals, hashtags, websites, or any kind of nodes in a network. Pause for a moment and think about what we are doing. Community detection is a clear, concise, and appropriate name for what we are doing. We are zooming in on communities that exist in a network. What communities would you be interested in exploring and understanding, and why?

There are many good use cases for this. You can use it to understand the sentiment communities share about your product. You can use this to understand a threat landscape. You can use this to understand how ideas move and transform between different groups of people. Be creative here. There are probably more uses for this than you can imagine.

In this chapter, we will explore this in the context of human life, but you should not feel limited to only using this for social network analysis. This is very useful in social network analysis, but it is also useful in analyzing most network data, not just social network data. For instance, this can be useful in both cybersecurity (malware analysis) and computational humanities, or in understanding how ideas move between groups and evolve.

There are at least three different approaches to doing community detection, with the most frequently researched including the following:

- Node connectivity

- Node closeness

- Network splitting

What I am calling *node connectivity* has to do with whether nodes are part of the same connected component or not. If two nodes are not part of the same connected component, then they are part of a completely different social group, not part of the same community.

Node closeness has to do with the distance between two nodes, even if they are part of the same connected component. For instance, two people might work together in the same large organization, but if they are more than two handshakes away from one another, they may not be part of the same community. It would take several rounds of introductions for them to ever meet each other. Consider how many people would you have to go through to be introduced to your favorite celebrity. How many people would you need to be introduced to?

Network splitting has to do with literally cutting a network into pieces by either removing nodes or edges. The preferred approach that I will explain is cuts on edges, but I have done something similar by removing nodes, and I've done this a few times in this book, shattering networks into pieces by removing central nodes.

I do not believe that we are at the end of discovery for community detection. I hope that reading through this chapter will give you some ideas for new approaches to identifying the various communities that exist in networks.

Getting started with community detection

Before we can start, we need a network to use. Let's use NetworkX's *Les Miserables* graph that we used in the previous chapter since it held several separate communities:

1. Loading the network is simple:

    ```
    import networkx as nx
    import pandas as pd
    G = nx.les_miserables_graph()
    ```

    That's all it takes to load the graph.

2. There is a `weight` attribute that I do not want to include in the network because we don't need edge weights for this simple demonstration. So, I'm going to drop it and rebuild the graph:

```
df = nx.to_pandas_edgelist(G)[['source', 'target']]
# dropping 'weight'
G = nx.from_pandas_edgelist(df)
```

In those two steps, we converted the *Les Miserables* graph into a `pandas` edge list, and we kept only the `source` and `target` fields, effectively dropping the `weight` field. Let's see how many nodes and edges exist in the network:

```
nx.info(G)
'Graph with 77 nodes and 254 edges'
```

This is a tiny network. Does this network contain isolates and islands, or is it just one large connected component? Let's check.

3. First, let's add the `draw_graph` function:

```
def draw_graph(G, show_names=False, node_size=1, font_
size=10, edge_width=0.5):

    import numpy as np
    from IPython.display import SVG
    from sknetwork.visualization import svg_graph
    from sknetwork.data import Bunch
    from sknetwork.ranking import PageRank

    adjacency = nx.to_scipy_sparse_matrix(G, nodelist=
None, dtype=None, weight='weight', format='csr')
    names = np.array(list(G.nodes()))

    graph = Bunch()
    graph.adjacency = adjacency
    graph.names = np.array(names)
    pagerank = PageRank()
    scores = pagerank.fit_transform(adjacency)

    if show_names:
        image = svg_graph(graph.adjacency, font_size
= font_size , node_size=node_size, names=graph.names,
```

```
width=700, height=500, scores=scores, edge_width = edge_
width)
    else:
        image = svg_graph(graph.adjacency, node_size =
node_size, width=700, height=500, scores = scores, edge_
width=edge_width)


    return SVG(image)
```

4. Now, let's visualize the network in its entirety:

```
draw_graph(G, font_size=12, show_names=True, node_size
=4, edge_width=1)
```

This outputs the following:

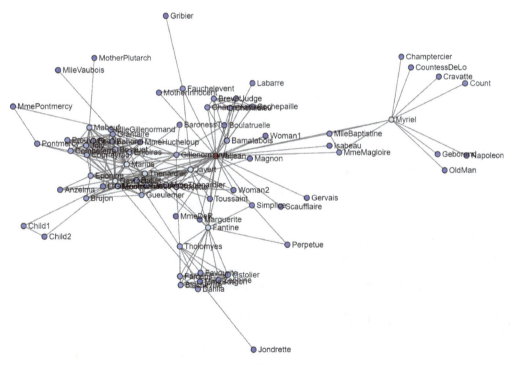

Figure 9.1 – Les Miserables graph

At a glance, we should be able to see that there are no isolates (nodes without edges), there are several nodes with a single edge, there are several clusters of nodes that are very close to each other (communities), and there are a few critically important nodes. If those critically important nodes were removed, the network would shatter to pieces.

5. Let's zoom in a little, using k_core, and only show nodes that have two or more edges. Let's also not display labels so that we can get a sense of the overall shape of the network:

```
draw_graph(nx.k_core(G, 2), font_size=12, show_
names=False, node_size=4, edge_width=0.5)
```

We will get the following output:

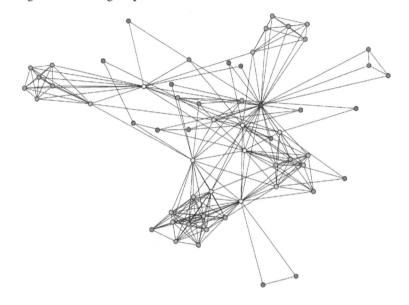

Figure 9.2 – Les Miserables graph, k_core with K=2, unlabeled

The communities should be a little clearer now. Look for parts of the graphs where the nodes are close together and where there are more edges/lines present. How many communities do you see? Four of them really stand out to me, but there are smaller groups scattered around, and there is also likely a community in the center of the network.

We are now ready to begin our attempts at community detection.

Exploring connected components

The first attempt at understanding the various communities and structures that exist in a network is often to analyze the connected components. As we discussed in *Chapter 7*, connected components are structures in networks where all nodes have a connection to another node in the same component.

As we saw previously, connected components can be useful for finding smaller connected components. Those can be thought of as communities as they are detached from the primary component and overall network, but the largest connected component is not typically a single community. It is usually made up of several communities, and it can usually be split into individual communities.

In the *Les Miserables* network, there is only one connected component. There are no islands or isolates. There is just one single component. That makes sense, as these are the characters from a piece of literature, and it wouldn't make much sense for characters in a book to just spend all day talking to themselves. However, that takes away a bit of the usefulness of inspecting connected components for this graph.

There is a way around that! As I mentioned previously, if we remove a few critically important nodes from a network, that network tends to shatter into pieces:

1. Let's remove five very important characters from the network:

    ```
    G_copy = G.copy()
    G_copy.remove_nodes_from(['Valjean', 'Marius', 'Fantine',
    'Cosette', 'Bamatabois'])
    ```

2. In these two lines, we built a second graph called `G_copy`, and then we removed five key nodes. Let's visualize the network again!

    ```
    draw_graph(G_copy, font_size=12, show_names=True, node_
    size=4, edge_width=1)
    ```

This gives us the following output:

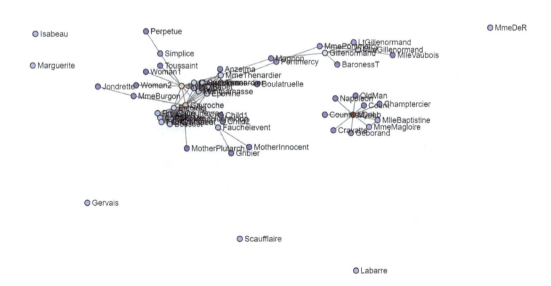

Figure 9.3 – Shattered Les Miserables network

Great. That's much closer to how many real-world networks look. There's still one primary connected component (continent), there are three smaller connected components (islands), and there are six isolate nodes. Calling these islands and continents is my own thing. There is no threshold for deciding that an island is a continent. It is just that most networks contain one super-component (continent), lots and lots of isolate nodes, and several connected components (islands). This helps me, but do with it what you like.

One other thing to keep in mind is that what we just did could be used as a step in community detection. Removing a few key nodes can break a network apart, pulling out the smaller communities that exist. Those critically important nodes held one or more communities together as part of the larger structure. Removing the important nodes allowed the communities to drift apart. We did this by removing important nodes, which is not usually ideal. However, other actual approaches to community detection work similarly, by removing edges rather than nodes.

3. How many connected components are left after shattering the network?

```
components = list(nx.connected_components(G_copy))
len(components)
10
```

NetworkX says that there are 10, but isolates are not connected to anything other than possibly themselves.

4. Let's remove them before looking into connected components:

```
G_copy = nx.k_core(G_copy, 1)
components = list(nx.connected_components(G_copy))
len(components)
4
```

It looks like there are four connected components.

5. As there are so few of them, let's inspect each one:

```
community = components[0]
G_community = G_copy.subgraph(community)
draw_graph(G_community, show_names=True, node_size=5)
```

Let's look at the visualization:

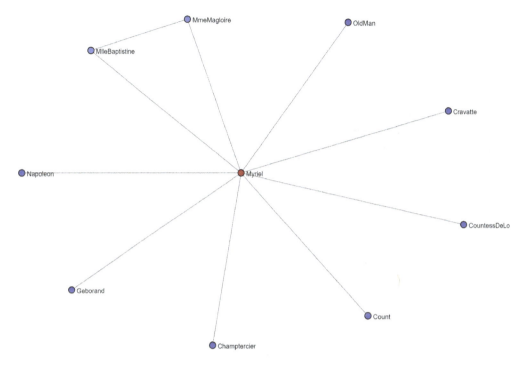

Figure 9.4 – Component 0 subgraph of the shattered Les Miserables network

Very interesting! The first connected component is almost a star network, with all nodes connecting to one central character, **Myriel**. However, if you look at the top left, you should see that two characters also share a link. That relationship could be worth investigating.

6. Let's look at the next component:

```
community = components[1]
G_community = G_copy.subgraph(community)
draw_graph(G_community, show_names=True, node_size=4)
```

This gives us the following output:

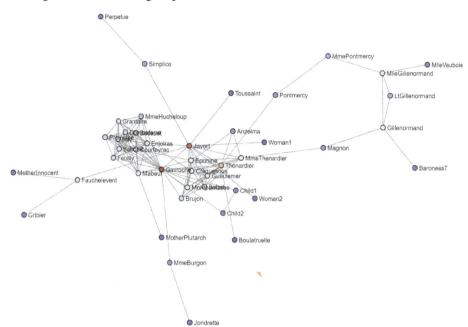

Figure 9.5 – Component 1 subgraph of the shattered Les Miserables network

This is even more interesting. This is what I am calling the primary component. It's the largest connected component in the shattered network. However, as I said, connected components are not ideal for identifying communities. Look slightly left of the center in the network – we should see two clusters of nodes, two separate communities. There's also at least one other community on the right. If two edges or nodes were removed, the community on the right would split off from the network. Onward!

7. Let's keep shattering the community:

```
community = components[2]
G_community = G_copy.subgraph(community)
draw_graph(G_community, show_names=True, node_size=4)
```

We will get the following output:

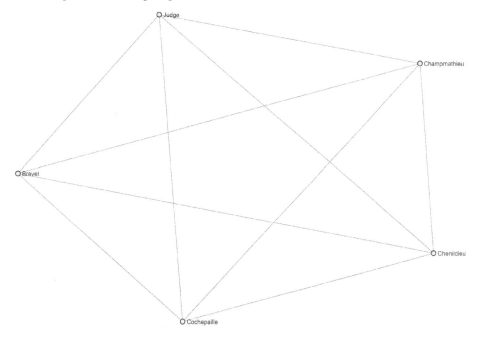

Figure 9.6 – Component 2 subgraph of the shattered Les Miserables network

This is a strongly connected component. Each node has a connection to the other nodes in this network. If one node were removed, this network would remain intact. From a network perspective, each node is as important or central as each other node.

8. Let's check the final component:

```
community = components[3]
G_community = G_copy.subgraph(community)
draw_graph(G_community, show_names=True, node_size=4)
```

This gives us the following visualization:

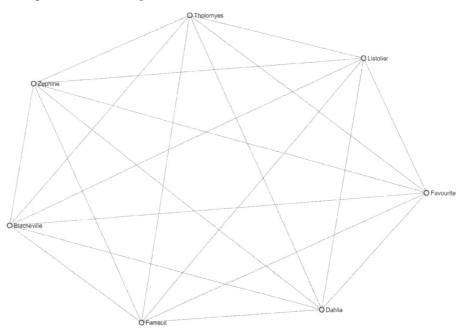

Figure 9.7 – Component 3 subgraph of the shattered Les Miserables network

This is another densely connected network. Each node is equally important or central. If one node were to be removed, this network would remain intact.

As you can see, we were able to find three communities by looking at the connected components, but connected components did not draw out the communities that exist in the larger primary component. If we wanted to draw those out, we'd have to remove other important nodes and then repeat our analysis. Throwing away nodes is one way to lose information, so I do not recommend that approach, but it can be useful for quick ad hoc analysis.

I do not consider investigating connected components to be community detection, but communities can be found while investigating connected components. I consider this one of the first steps that should be done during any network analysis, and the insights gained are valuable, but it's not sensitive enough for community detection.

If your network contained no super-cluster of a connected component, then connected components would be pretty adequate for community detection. However, you would have to treat the super-cluster as one community, and in reality, the cluster contains many communities. The connected component approach becomes less useful with larger networks.

Let's move on to more suitable methods.

Using the Louvain method

The **Louvain method** is certainly my favorite for community detection, for a few reasons.

First, this algorithm can be used on very large networks of millions of nodes and it will be effective and fast. Other approaches that we will explore in this chapter will not work on large networks and will not be as fast, so we get effectiveness and speed with this algorithm that we can't find anywhere else. As such, it is my go-to algorithm for community detection, and I save the others as options to consider.

Second, it is possible to tune the `resolution` parameter to find the best partitions for community detection, giving flexibility when the default results are not optimal. With the other algorithms, you do not have this flexibility.

In summary, with the Louvain method, we have a fast algorithm that is effective at community detection in massive networks, and we can optimize the algorithm for better results. I recommend dabbling in community detection by starting with the Louvain method, and then picking up these other approaches as you learn. It's good to know that there are options.

How does it work?

The creators of the Louvain method were able to use their algorithm on a network of hundreds of millions of nodes and more than a billion edges, making this approach suitable for very large networks. You can read more about the Louvain method at `https://arxiv.org/pdf/0803.0476.pdf`.

The algorithm works through a series of passes, where each pass contains two phases. The first phase assigns different communities to each node in the network. Initially, each node has a different community assigned to it. Then, each neighbor is evaluated and nodes are assigned to communities. The first step concludes when no more improvements can be made. In the second phase, a new network is built, with nodes being the communities discovered in the first step. Then, the results of the first phase can be repeated. The two steps are iterated until optimal communities are found.

This is a simplified description of how the algorithm works. I recommend reading the research paper in its entirety to get a feel for how the algorithm works.

The Louvain method in action!

We used the Louvain method briefly in *Chapter 3*, so if you paid attention, this code should look familiar. The Louvain method has been included in more recent versions of NetworkX, so if you have the latest version of NetworkX, you will not need to use the `community` Python library, but your code will be different. For consistency, I will use the "community" library approach:

1. First, let's import the library:

    ```
    import community as community_louvain
    ```

2. Here is some code that will help us draw Louvain partitions:

```
def draw_partition(G, partition):

    import matplotlib.cm as cm
    import matplotlib.pyplot as plt

    # draw the graph
    plt.figure(3,figsize=(12,12))
    pos = nx.spring_layout(G)

    # color the nodes according to their partition
    cmap = cm.get_cmap('jet', max(partition.values()) +
1)
    nx.draw_networkx_nodes(G, pos, partition.keys(),
node_size=40, cmap=cmap, node_color = list(partition.
values()))
    nx.draw_networkx_edges(G, pos, alpha=0.5, width =
0.3)

    return plt.show()
```

3. Now, let's use the best_partition function to identify the optimal partition using the Louvain method. During my testing, I found resolution=1 to be ideal, but with other networks, you should experiment with this parameter:

```
partition = community_louvain.best_partition(G,
resolution=1)
draw_partition(G, partition)
```

This creates a visualization:

Figure 9.8 – Louvain method community detection of the Les Miserables network

The helper function in *step 2* will color nodes by the communities that they belong to. What is important is that the separate communities have been detected, and each community of nodes is identified with a different color. Each node belongs to a different partition, and those partitions are the communities.

4. Let's take a look at what is inside the `partition` variable:

```
partition
{'Napoleon': 1,
 'Myriel': 1,
 'MlleBaptistine': 1,
 'MmeMagloire': 1,
 'CountessDeLo': 1,
 'Geborand': 1,

 ...

 'Grantaire': 0,
 'Child1': 0,
 'Child2': 0,
```

```
'BaronessT': 2,
'MlleVaubois': 2,
'MotherPlutarch': 0}
```

To save space, I cut out some of the nodes and partitions. Each node has an associated partition number, and that's the community that it belongs to. If you wanted to get a list of nodes that belong to an individual community, you could do something like this:

```
[node for node, community in partition.items() if
community == 2]
```

So, why is this exciting? What's so cool about the Louvain method? Well, for one thing, it can scale to massive networks, allowing for research into the largest networks, such as the internet. Second, it's fast, which means it is practical. There is not a lot of point to an algorithm that is so slow as to only be useful on tiny networks. Louvain is practical with massive networks. This algorithm is fast and efficient, and the results are very good. This is one algorithm for community detection that you will want in your tool belt.

Next, let's look at label propagation as another option for community detection.

Using label propagation

Label propagation is another fast approach for identifying communities that exist in a network. In my experience, the results haven't been as good as with the Louvain method, but this is another tool that can be explored as part of community detection. You can read about label propagation at https://arxiv.org/pdf/0709.2938.pdf.

How does it work?

This is an iterative approach. Each node is initialized with a unique label, and during each iteration of the algorithm, each node adopts the label that most of its neighbors have. For instance, if the **David** node had seven neighbor nodes, and four out of seven neighbors were **label 1** with the other three were **label 0**, then the **David** node would pick up **label 1**. During each step of the process, each node picks up the majority label, and the process concludes by grouping nodes with the same labels together as communities.

Label propagation in action!

This algorithm can be imported directly from NetworkX:

```
from networkx.algorithms.community.label_propagation import
label_propagation_communities
```

Once you have imported the algorithm, all you have to do is pass it to your graph, and you will get back a list of communities:

1. Let's try this out using our Les Miserables graph:

    ```
    communities = label_propagation_communities(G)
    ```

    This line passes our graph to the label propagation algorithm and writes the results to a `community` variable. On a network this small, this algorithm is zippy-fast, taking a fraction of a second to identify communities. I prefer to convert these results into a list, to extract the community nodes.

2. We can do just that, like so:

    ```
    communities = list(communities)
    communities[0]
    {'Champtercier',
     'Count',
     'CountessDeLo',
     'Cravatte',
     'Geborand',
     'Myriel',
     'Napoleon',
     'OldMan'}
    ```

    In that last line, we inspected the first community, community 0. Visualizing these communities is simple enough.

3. We can extract them as subgraphs and then use the same `draw_graph` function we have used throughout this book:

    ```
    community = communities[1]
    G_community = G.subgraph(community)
    draw_graph(G_community, show_names=True, node_size=5)
    ```

What can we see from the output?

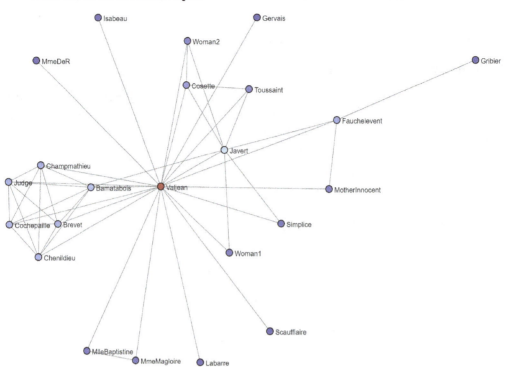

Figure 9.9 – Label propagation community detection of the Les Miserables network, community 1

This looks pretty good, but not quite as good as the results from the Louvain method. It was fast but not quite as precise as I'd wanted. For instance, looking to the left of **Valjean**, there is a tight community of densely connected nodes. That should be its own community, not part of this larger group. This algorithm isn't perfect, but no algorithm is. However, this algorithm is fast and can scale to large networks, so it is another option for large-scale community detection.

4. Let's look at a couple more communities:

```
community = communities[2]
G_community = G.subgraph(community)
draw_graph(G_community, show_names=True, node_size=5)
```

This gives us the following output:

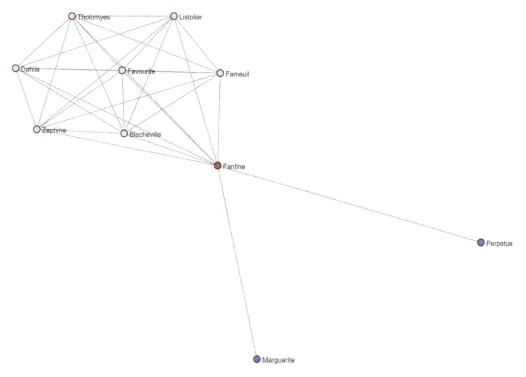

Figure 9.10 – Label propagation community detection of the Les Miserables network, community 2

This community looks about perfect. It's not uncommon to have a few additional nodes other than those found in the most densely connected parts of a community.

5. Let's look at another:

```
community = communities[3]
G_community = G.subgraph(community)
draw_graph(G_community, show_names=True, node_size=5)
```

This gives us the following visualization:

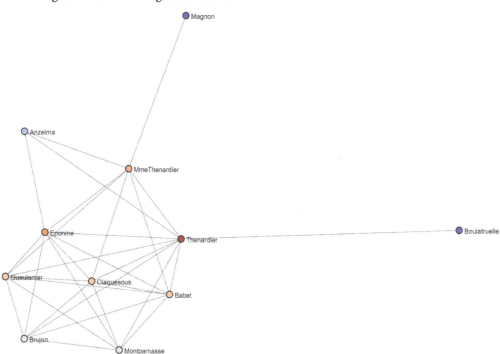

Figure 9.11 – Label propagation community detection of the Les Miserables network, community 3

This community looks great as well. Overall, this algorithm works well and is fast. Also, the setup is easier and faster than with the Louvain method, as all you have to do is import the algorithm, pass it a graph, and then visualize the results. In terms of ease of use, this algorithm is the easiest that I've seen. The results look good, and communities have quickly been identified.

But being fast and easy to use is not enough. Louvain is more accurate and is fast and easy to use. Still, this can be useful.

Using the Girvan-Newman algorithm

At the beginning of this chapter, we noticed that the *Les Miserables* network consisted of a single large connected component and that there were no isolates or smaller "islands" of communities apart from the large connected component. To show how connected components could be useful for identifying communities, we shattered the network by removing a few key nodes.

That approach is not typically ideal. While there is information in both nodes (people, places, things) and edges (relationships), in my experience, it is typically preferable to throw away edges than to throw away nodes.

A better approach than what we did previously would be to identify the least number of edges that could be cut that would result in a split network. We could do this by looking for the edges that the greatest number of shortest paths pass through – that is, the edges with the highest `edge_betweenness_centrality`.

That is precisely what the **Girvan-Newman algorithm** does.

How does it work?

The Girvan-Newman algorithm identifies communities by cutting the least number of edges possible, which results in splitting a network into two pieces. You can read more about their approach here: `https://www.pnas.org/doi/full/10.1073/pnas.122653799`.

Many times, when I'm looking at networks, I see several nodes on two different sides connected by a few edges. It almost looks like a few rubber bands are holding the two groups together. If you snip the rubber bands, the two communities should fly apart, similar to how networks shatter into pieces when key nodes are removed.

In a way, this is more surgically precise than removing nodes. There is less loss of valuable information. Sure, losing information on certain relationships is a drawback, but all of the nodes remain intact.

Through a series of iterations, the Girvan-Newman algorithm identifies edges with the highest `edge_betweenness_centrality` scores and removes them, splitting a network into two pieces. Then, the process begins again. If not repeated enough, communities are too large. If repeated too many times, communities end up being a single node. So, there will be some experimentation when using this algorithm to find the ideal number of cuts.

This algorithm is all about cutting. The downside of this algorithm is that it is not fast. Calculating `edge_betweenness_centrality` is much more computationally expensive than the computations being done for the Louvain method or label propagation. As a result, this algorithm ceases to be useful very quickly, as it becomes much too slow to be practical.

However, if your network is small enough, this is a very cool algorithm to explore for community detection. It's also intuitive and easy to explain to others.

Girvan-Newman algorithm in action!

Let's try this out with our *Les Miserables* graph. The graph is small enough that this algorithm should be able to split it into communities pretty quickly:

1. First, import the algorithm:

   ```
   from networkx.algorithms.community import girvan_newman
   ```

2. Next, we need to pass the graph to the algorithm as a parameter. When we do this, the algorithm will return the results of each iteration of splits, which we can investigate by converting the results into a list:

   ```
   communities = girvan_newman(G)
   communities = list(communities)
   ```

3. What was the maximum number of iterations that the algorithm could do before each community consisted of a single node? Let's see:

   ```
   len(communities)
   76
   ```

   Neat! We have 76 iterations of splits kept in a Python list. I recommend that you investigate the various levels of splits and find the one that looks best for your needs. It could be very early in the process, in the first 10 splits, or it might be a bit later. This part requires some analysis, further making this a bit of a hands-on algorithm.

4. However, just to push forward, let's pretend that we found that the tenth iteration of splits yielded the best results. Let's set the tenth iteration results as our final group of communities, and then visualize the communities as we did with the Louvain method and label propagation:

   ```
   communities = communities[9]
   ```

   We're keeping the tenth iteration results and dropping everything else. If we didn't want to throw away the results, we could have used a different variable name.

5. Let's see what these communities look like so that we can compare them against the other algorithms we discussed:

   ```
   community = communities[0]
   G_community = G.subgraph(community)
   draw_graph(G_community, show_names=True, node_size=5)
   ```

We get the following output:

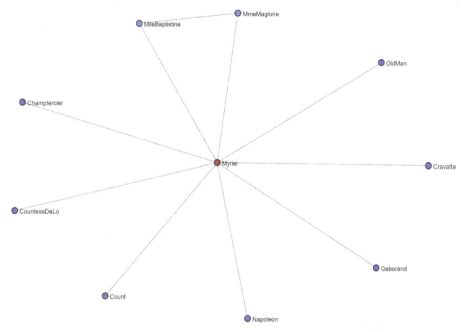

Figure 9.12 – Girvan-Newman community detection of the Les Miserables network, community 0

This subgraph should look familiar! We saw exactly this when we shattered the network by nodes and then visualized connected components. This algorithm split the network using edges and managed to find the same community.

6. Let's see another community:

```
community = communities[1]
G_community = G.subgraph(community)
draw_graph(G_community, show_names=True, node_size=5)
```

This produces the following network visualization:

Figure 9.13 – Girvan-Newman community detection of the Les Miserables network, community 1

This also looks very good. It's not uncommon for communities to have a densely connected group, as well as some less connected nodes.

7. And another community:

```
community = communities[2]
G_community = G.subgraph(community)
draw_graph(G_community, show_names=True, node_size=5)
```

We will see the following output:

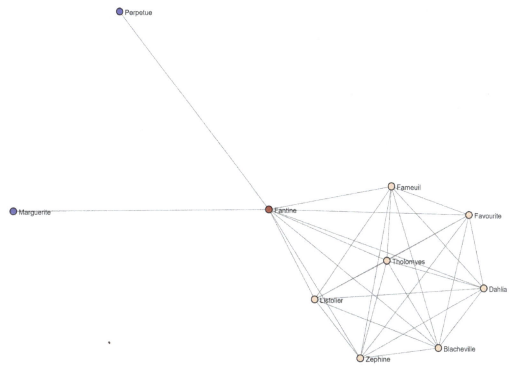

Figure 9.14 – Girvan-Newman community detection of the Les Miserables network, community 2

This is similar to the last community. We have a densely connected group of nodes and two nodes with a single edge. This looks great.

8. Let's see community 3:

```
community = communities[3]
G_community = G.subgraph(community)
draw_graph(G_community, show_names=True, node_size=5)
```

Community 3 looks like this:

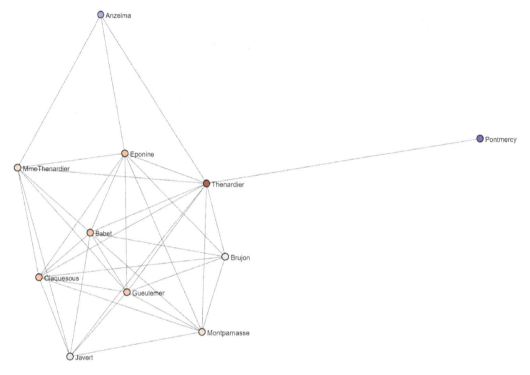

Figure 9.15 – Girvan-Newman community detection of the Les Miserables network, community 3

This should look familiar as well. Label propagation found the same community but Girvan-Newman removed one additional node.

9.   And the next one:

```
community = communities[4]
G_community = G.subgraph(community)
draw_graph(G_community, show_names=True, node_size=5)
```

We will see the following network:

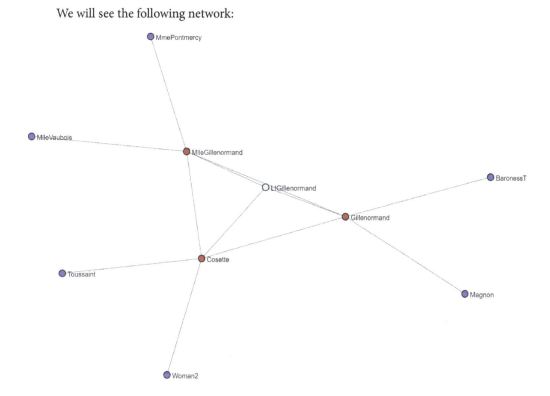

Figure 9.16 – Girvan-Newman community detection of the Les Miserables network, community 4

While this may be less visually appealing to look at, this impresses me more than the other network visualizations. This is a less obvious community, found only by cutting edges with the highest edge_ betweenness_centrality scores. There is a slightly more connected group of nodes in the center, surrounded by nodes with a single edge each on the outskirts.

The Girvan-Newman algorithm can give really good and clean results. The only downside is its speed. Calculating edge_betweenness_centrality and shortest_paths is time-consuming, so this algorithm is much slower than the others that we discussed, but it can be very useful if your network is not too large.

Other approaches to community detection

All of these algorithms that we have explored were ideas that people had on how to identify communities in networks, either based on nearness to other nodes or found by cutting edges. However, these are not the only approaches. I came up with an approach before learning about the Girvan-Newman algorithm that cut nodes rather than edges. However, when I learned about the Girvan-Newman approach, I found that to be more ideal and gave up on my implementation. But that makes me think, what other approaches might there be for identifying communities in networks?

As you learn more and become more comfortable working with networks, try to discover other ways of identifying communities.

Summary

In this chapter, we went through several different approaches to community detection. Each had its pros and cons.

We saw that connected components can be useful for identifying communities, but only if the network consists of more than just one single primary component. To use connected components to identify communities, there need to be some smaller connected components split off. It's very important to use connected components at the beginning of your network analysis to get an understanding of the overall structure of your network, but it is less than ideal as a standalone tool for identifying communities.

Next, we used the Louvain method. This algorithm is extremely fast and can be useful in networks where there are hundreds of millions of nodes and billions of edges. If your network is very large, this is a useful first approach for community detection. The algorithm is fast, and the results are clean. There is also a parameter you can experiment with to get optimal partitions.

We then used label propagation to identify communities. On the *Les Miserables* network, the algorithm took a fraction of a second to identify communities. Overall, the results looked good, but it did seem to struggle with splitting out a dense cluster of nodes from a larger community. However, every other community looked good. This algorithm is fast and should scale to large networks, but I have never heard of this being used on a network with millions of nodes. It is worth experimenting with.

Finally, we used the Girvan-Newman algorithm, which is an algorithm that finds communities by performing several rounds of cuts on edges with the highest `edge_betweenness_centrality` scores. The results were very clean. The downside of this algorithm is that it is very slow and does not scale well to large networks. However, if your network is small, then this is a very useful algorithm for community detection.

This has been a fun chapter to write. Community detection is one of the most interesting areas of network analysis, for me. It's one thing to analyze networks as a whole or explore ego networks, but being able to identify and extract communities is another skill that sits somewhere between whole network analysis and egocentric network analysis.

In the next few chapters, we're going to go into the wilderness and explore how we can use network science and machine learning together! The first chapter will be on supervised machine learning, while the final chapter will be on unsupervised machine learning. We only have a few more chapters to go! Hang in there!

10

Supervised Machine Learning on Network Data

In previous chapters, we spent a lot of time exploring how to collect text data from the internet, transform it into network data, visualize networks, and analyze networks. We were able to use centralities and various network metrics for additional contextual awareness about individual nodes' placement and influence in networks, and we used community detection algorithms to identify the various communities that exist in a network.

In this chapter, we are going to begin an exploration of how network data can be useful in **machine learning (ML)**. As this is a data science and network science book, I expect that many readers will be familiar with ML, but I'll give a very quick explanation.

This chapter is composed of the following sections:

- Introducing ML
- Beginning with ML
- Data preparation and feature engineering
- Selecting a model
- Preparing the data
- Training and validating the model
- Model insights
- Other use cases

Technical requirements

In this chapter, we will be using the Python libraries NetworkX, pandas, and scikit-learn. These libraries should be installed by now, so they should be ready for your use. If they are not installed, you can install Python libraries with the following:

```
pip install <library name>
```

For instance, to install NetworkX, you would do this:

```
pip install networkx
```

In *Chapter 4*, we also introduced a `draw_graph()` function that uses both NetworkX and `scikit-network`. You will need that code any time that we do network visualization. Keep it handy!

The code for this chapter is on GitHub: `https://github.com/PacktPublishing/Network-Science-with-Python`.

Introducing ML

ML is a set of techniques that enable computers to learn from patterns and behavior in data. It is often said that there are three different kinds of ML: **Supervised**, **Unsupervised**, and **Reinforcement** learning.

In supervised ML, an answer – called a **label** – is provided with the data to allow for an ML model to learn the patterns that will allow it to predict the correct answer. To put it simply, you give the model data *and* an answer, and it figures out how to predict correctly.

In unsupervised ML, no answer is provided to the model. The goal is usually to find clusters of similar pieces of data. For instance, you could use clustering to identify the different types of news articles present in a dataset of news articles, or to find topics that exist in a corpus of text. This is similar to what we have done with community detection.

In reinforcement learning, a model is given a goal and it gradually learns how to get to this goal. In many reinforcement learning demos, you'll see a model play pong or another video game, or learn to walk.

These are ultra-simplistic descriptions of the types of ML, and there are more variations (**semi-supervised** and so on). ML is a rich topic, so I encourage you to find books if this chapter interests you. For me, it pushed NLP into an obsession.

Beginning with ML

There are many guides and books on how to do sentiment analysis using NLP. There are very few guides and books that give a step-by-step demonstration of how to convert graph data into a format that can be used for classification with ML. In this chapter, you will see how to use graph data for ML classification.

For this exercise, I created a little game I'm calling "Spot the Revolutionary." As with the last two chapters, we will be using the networkx *Les Miserables* network as it contains enough nodes and communities to be interesting. In previous chapters, I pointed out that the revolutionary community was densely connected. As a reminder, this is what it looks like:

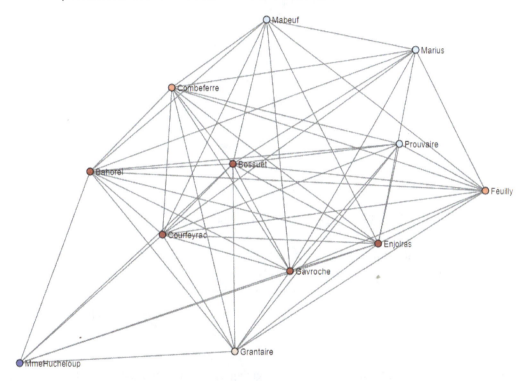

Figure 10.1 – Les Miserables Les Amis de l'ABC network

Each member of the community is connected with each other member, for the most part. There are no connections with outside people.

Other parts of the network look different.

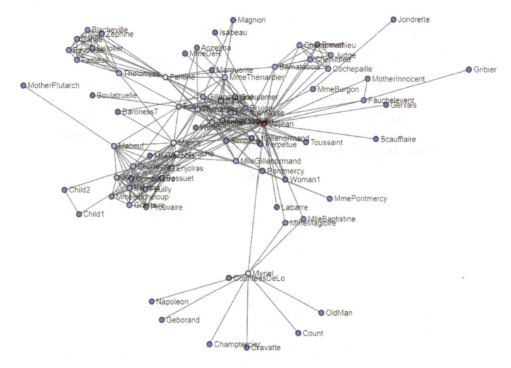

Figure 10.2 – Les Miserables whole network

Even a visual inspection shows that in different parts of the network, nodes are are conncteded differently. The structure is different. In some places, connectivity resembles a star. In others, it resembles a mesh. Network metrics will give us these values, and ML models will be able to use them for prediction.

We are going to use these metrics to play a game of "Spot the Revolutionary." This will be fun.

> **Note**
>
> I will not be explaining ML in much depth, only giving a preview of its capabilities. If you are interested in data science or software engineering, I strongly recommend that you spend some time learning about ML. It is not only for academics, mathematicians, and scientists. It gets complicated, so a foundation in mathematics and statistics is strongly recommended, but you should feel free to explore the topic.

This chapter will not be a mathematics lesson. All work will be done via code. I'm going to show one example of using network data for classification. This is not the only use case. This is not at all the only use case for using network data with ML. I will also be showing only one model (Random Forest), not all available models.

I'm also going to show that sometimes incorrect predictions can be as insightful as correct ones and how there are sometimes useful insights in model predictions.

I'm going to show the workflow to go from graph data to prediction and insights so that you can use this in your own experiments. You don't need a graph **neural network** (**NN**) for everything. This is totally possible with simpler models, and they can give good insights, too.

Enough disclaimers. Let's go.

Data preparation and feature engineering

Before we can use ML, we first need to collect our data and convert it into a format that the model can use. We can't just feed the graph G to Random Forest and call it a day. We could feed a graph's adjacency matrix and a set of labels to Random Forest and it'd work, but I want to showcase some feature engineering that we can do.

Feature engineering is using domain knowledge to create additional features (most call them columns) that will be useful for our models. For instance, looking at the networks from the previous section, if we want to be able to spot the revolutionaries, then we may want to give our model additional data such as each node's number of degrees (connections), betweenness centrality, closeness centrality, page rank, clustering, and triangles:

1. Let's start by first building our network. This should be easy by now, as we have done this several times:

    ```
    import networkx as nx
    import pandas as pd
    G = nx.les_miserables_graph()
    df = nx.to_pandas_edgelist(G)[['source', 'target']] #
    dropping 'weight'
    G = nx.from_pandas_edgelist(df)
    ```

    We took these steps in previous chapters, but as a reminder, the *Les Miserables* graph comes with edge weights, and I don't want those. The first line loads the graph. The second line creates an edge list from the graph, dropping the edge weights. And the third line rebuilds the graph from the edge list, without weights.

2. We should now have a useful graph. Let's take a look:

    ```
    draw_graph(G)
    ```

That produces the following graph:

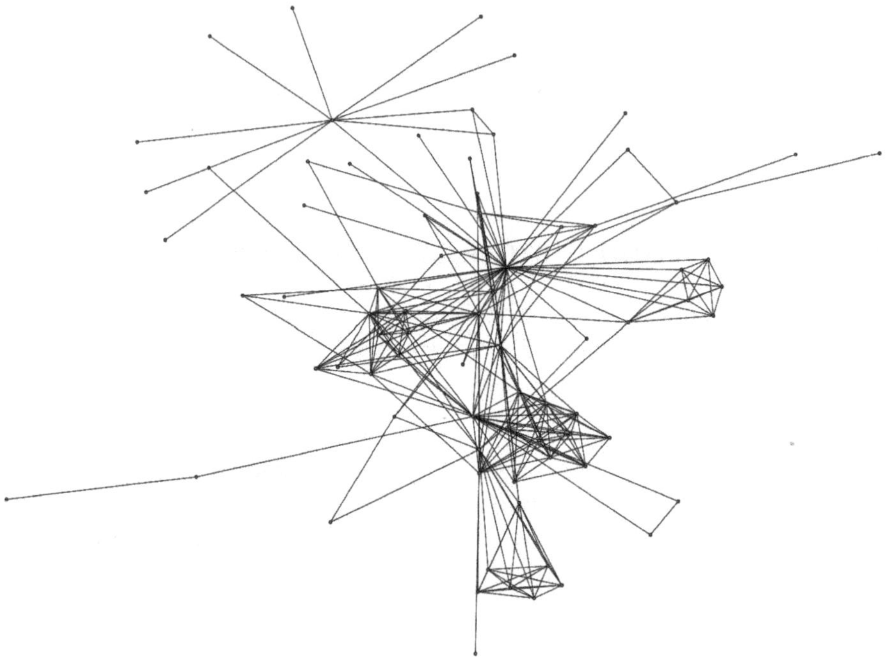

Figure 10.3 – Les Miserables graph without node names

Looks good! I can clearly see that there are several different clusters of nodes in this network and that other parts of the network are more sparse.

But how do we convert this crazy tangled knot into something that an ML model can use? Well, we looked at centralities and other measures in previous chapters. So, we already have the foundation that we will use here. I'm going to create several DataFrames with the data that I want and then merge them together for use as training data.

Degrees

The **degrees** are simply the number of connections that a node has with other nodes. We'll grab that first:

```
degree_df = pd.DataFrame(G.degree)
degree_df.columns = ['person', 'degrees']
degree_df.set_index('person', inplace=True)
degree_df.head()
```

We get the following output:

person	degrees
Napoleon	1
Myriel	10
MlleBaptistine	3
MmeMagloire	3
CountessDeLo	1

Figure 10.4 – Les Miserables feature engineering: degrees

Let's move on to the next step.

Clustering

Next, we'll grab the clustering coefficient, which tells us how densely connected the nodes are around a given node. A value of 1.0 means that every node is connected to every other node. 0.0 means that no neighbor nodes are connected to other neighbor nodes.

Let's capture clustering:

```
clustering_df = pd.DataFrame(nx.clustering(G), index=[0]).T

clustering_df.columns = ['clustering']

clustering_df.head()
```

This gives us the clustering output:

	clustering
Napoleon	0.000000
Myriel	0.066667
MlleBaptistine	1.000000
MmeMagloire	1.000000
CountessDeLo	0.000000

Figure 10.5 – Les Miserables feature engineering: clustering

This tells us that **MlleBaptistine** and **MmeMagloire** both are part of densely connected communities, meaning that these two also know the same people. **Napoleon** doesn't have any overlap with other people, and neither does **CountessDeLo**.

Triangles

Triangles are similar to clustering. Clustering has to do with how many triangles are found around a given node compared to the number of possible triangles. `triangle_df` is a count of how many triangles a given node is a part of. If a node is part of many different triangles, it is connected with many nodes in a network:

```
triangle_df = pd.DataFrame(nx.triangles(G), index=[0]).T
triangle_df.columns = ['triangles']
triangle_df.head()
```

This gives us the following:

	triangles
Napoleon	0
Myriel	3
MlleBaptistine	3
MmeMagloire	3
CountessDeLo	0

Figure 10.6 – Les Miserables feature engineering: triangles

This is another way of understanding the connectedness of the nodes around a node. These nodes are people, so it is a way of understanding the connectedness of peopleto other people. Notice that the results are similar but not identical to clustering.

Betweenness centrality

Betweenness centrality has to do with a node's placement between other nodes. As a reminder, in a hypothetical situation where there are three people (*A*, *B*, and *C*), and if *B* sits between *A* and *C*, then all information passing from *A* to *C* flows through person *B*, putting them in an important and influential position. That's just one example of the usefulness of betweenness centrality. We can get this information by using the following code:

```
betw_cent_df = pd.DataFrame(nx.betweenness_centrality(G),
index=[0]).T
```

```
betw_cent_df.columns = ['betw_cent']
betw_cent_df.head()
```

This gives us the following output:

	betw_cent
Napoleon	0.000000
Myriel	0.176842
MlleBaptistine	0.000000
MmeMagloire	0.000000
CountessDeLo	0.000000

Figure 10.7 – Les Miserables feature engineering: betweenness centrality

Closeness centrality

Closeness centrality has to do with how close a given node is to all other nodes in a network. It has to do with the shortest path. As such, closeness centrality is computationally very slow for large networks. However, it will work just fine for the *Les Miserables* network, as this is a very small network:

```
close_cent_df = pd.DataFrame(nx.closeness_centrality(G),
index=[0]).T
close_cent_df.columns = ['close_cent']
close_cent_df.head()
```

	close_cent
Napoleon	0.301587
Myriel	0.429379
MlleBaptistine	0.413043
MmeMagloire	0.413043
CountessDeLo	0.301587

Figure 10.8 – Les Miserables feature engineering: closeness centrality

PageRank

Finally, the **PageRank** algorithm was created by the founders of Google and is similar to other centrality measures in that it is useful for gauging the importance of a node in a network. Also, as betweenness centrality and closeness centrality become impractically slow as networks become large, `pagerank` remains useful even on large networks. As such, it is very commonly used to gauge importance:

```
pr_df = pd.DataFrame(nx.pagerank(G), index=[0]).T
pr_df.columns = ['pagerank']
pr_df.head()
```

This gives us *Figure 10.9*.

	pagerank
Napoleon	0.005584
Myriel	0.042803
MlleBaptistine	0.010279
MmeMagloire	0.010279
CountessDeLo	0.005584

Figure 10.9 – Les Miserables feature engineering: pagerank

Adjacency matrix

Finally, we can include the **adjacency matrix** in our training data so that our models can use neighbor nodes as features for making predictions. For instance, let's say that you have 10 friends but one of them is a criminal, and every person that friend introduces you to is also a criminal. You will probably learn over time that you shouldn't associate with that friend or their friends. Your other friends do not have that problem. In your head, you've already begun to make judgments about that person and who they associate with.

If we left the adjacency matrix out, the model would attempt to learn from the other features, but it'd have no contextual awareness of neighboring nodes. In the game of "Spot the Revolutionary," it would use the centralities, clustering, degrees, and other features only, as it'd have no way of learning from anything else.

We're going to use the adjacency matrix. It feels almost like leakage (where the answer is hidden in another feature) because like often attracts like in a social network, but that also shows the usefulness of using network data with ML. You can drop the adjacency matrix if you feel that it is cheating. I do not:

```
adj_df = nx.to_pandas_adjacency(G)
adj_df.columns = ['adj_' + c.lower() for c in adj_df.columns]
adj_df.head()
```

This code outputs the following DataFrame:

	adj_napoleon	adj_myriel	adj_mllebaptistine	adj_mmemagloire	adj_countessdelo	adj_geborand	adj_champtercier	adj_cravatte	adj_count
Napoleon	0.0	1.0	0.0	0.0	0.0	0.0	0.0	0.0	0.0
Myriel	1.0	0.0	1.0	1.0	1.0	1.0	1.0	1.0	1.0
MlleBaptistine	0.0	1.0	0.0	1.0	0.0	0.0	0.0	0.0	0.0
MmeMagloire	0.0	1.0	1.0	0.0	0.0	0.0	0.0	0.0	0.0
CountessDeLo	0.0	1.0	0.0	0.0	0.0	0.0	0.0	0.0	0.0
Geborand	0.0	1.0	0.0	0.0	0.0	0.0	0.0	0.0	0.0
Champtercier	0.0	1.0	0.0	0.0	0.0	0.0	0.0	0.0	0.0
Cravatte	0.0	1.0	0.0	0.0	0.0	0.0	0.0	0.0	0.0
Count	0.0	1.0	0.0	0.0	0.0	0.0	0.0	0.0	0.0
OldMan	0.0	1.0	0.0	0.0	0.0	0.0	0.0	0.0	0.0

10 rows × 77 columns

Figure 10.10 – Les Miserables feature engineering: adjacency matrix

Merging DataFrames

Now that we have all of these useful features, it's time to merge the DataFrames together. This is simple, but requires a few steps, as demonstrated in the following code:

```
clf_df = pd.DataFrame()
clf_df = degree_df.merge(clustering_df, left_index=True, right_index=True)
clf_df = clf_df.merge(triangle_df, left_index=True, right_index=True)
clf_df = clf_df.merge(betw_cent_df, left_index=True, right_index=True)
clf_df = clf_df.merge(close_cent_df, left_index=True, right_index=True)
clf_df = clf_df.merge(pr_df, left_index=True, right_index=True)
clf_df = clf_df.merge(adj_df, left_index=True, right_index=True)
clf_df.head(10)
```

In the first step, I created an empty DataFrame just so that I could rerun the Jupyter cell over and over without creating duplicate columns with weird names. It just saves work and aggravation. Then, I sequentially merge the DataFrames into `clf_df`, based on the DataFrame indexes. The DataFrame indexes are character names from Les Miserables. This just makes sure that each row from each DataFrame is joined together correctly.

	degrees	clustering	triangles	betw_cent	close_cent	pagerank	adj_napoleon	adj_myriel	adj_mllebaptistine	adj_mmemagloire	...
Napoleon	1	0.000000	0	0.000000	0.301587	0.005584	0.0	1.0	0.0	0.0	...
Myriel	10	0.066667	3	0.176842	0.429379	0.042803	1.0	0.0	1.0	1.0	...
MlleBaptistine	3	1.000000	3	0.000000	0.413043	0.010279	0.0	1.0	0.0	1.0	...
MmeMagloire	3	1.000000	3	0.000000	0.413043	0.010279	0.0	1.0	1.0	0.0	...
CountessDeLo	1	0.000000	0	0.000000	0.301587	0.005584	0.0	1.0	0.0	0.0	...
Geborand	1	0.000000	0	0.000000	0.301587	0.005584	0.0	1.0	0.0	0.0	...
Champtercier	1	0.000000	0	0.000000	0.301587	0.005584	0.0	1.0	0.0	0.0	...
Cravatte	1	0.000000	0	0.000000	0.301587	0.005584	0.0	1.0	0.0	0.0	...
Count	1	0.000000	0	0.000000	0.301587	0.005584	0.0	1.0	0.0	0.0	...
OldMan	1	0.000000	0	0.000000	0.301587	0.005584	0.0	1.0	0.0	0.0	...

10 rows × 83 columns

Figure 10.11 – Les Miserables feature engineering: combined training data

Adding labels

Finally, we need to add labels for the revolutionaries. I have done the work of quickly looking up the names of the members of Les Amis de l'ABC (Friends of the ABC), which is the name of the group of revolutionaries. First, I will add the members to a Python list, and then I'll do a spot check to make sure that I've spelled their names correctly:

```
revolutionaries = ['Bossuet', 'Enjolras', 'Bahorel',
'Gavroche', 'Grantaire',
                    'Prouvaire', 'Courfeyrac', 'Feuilly',
'Mabeuf', 'Marius', 'Combeferre']
# spot check
clf_df[clf_df.index.isin(revolutionaries)]
```

This produces the following DataFrame:

	degrees	clustering	triangles	betw_cent	close_cent	pagerank	adj_napoleon	adj_myriel	adj_mllebaptistine	adj_mmemagloire	...
Gavroche	22	0.354978	82	0.165113	0.513514	0.035764	0.0	0.0	0.0	0.0	...
Marius	19	0.333333	57	0.132032	0.531469	0.030893	0.0	0.0	0.0	0.0	...
Enjolras	15	0.609524	64	0.042553	0.481013	0.021880	0.0	0.0	0.0	0.0	...
Bossuet	13	0.769231	60	0.030754	0.475000	0.018957	0.0	0.0	0.0	0.0	...
Mabeuf	11	0.690909	38	0.027661	0.395833	0.017476	0.0	0.0	0.0	0.0	...
Courfeyrac	13	0.756410	59	0.005267	0.400000	0.018576	0.0	0.0	0.0	0.0	...
Combeferre	11	0.927273	51	0.001250	0.391753	0.015890	0.0	0.0	0.0	0.0	...
Prouvaire	9	1.000000	36	0.000000	0.356808	0.013144	0.0	0.0	0.0	0.0	...
Feuilly	11	0.927273	51	0.001250	0.391753	0.015890	0.0	0.0	0.0	0.0	...
Bahorel	12	0.863636	57	0.002185	0.393782	0.017197	0.0	0.0	0.0	0.0	...
Grantaire	10	0.933333	42	0.000150	0.358491	0.014455	0.0	0.0	0.0	0.0	...

11 rows × 83 columns

Figure 10.12 – Les Misérables feature engineering: Friends of the ABC members

This looks perfect. The list had 11 names, and the DataFrame has 11 rows. To create training data for supervised ML, we need to add a **label** field so that the model will be able to learn to predict an answer correctly. We're going to use the same Python list as previously, and we're going to give each of these members a label of 1:

```
clf_df['label'] = clf_df.index.isin(revolutionaries).
astype(int)
```

It's that easy. Let's take a quick look at the DataFrame just to make sure we have labels. I'll sort on the index so that we can see a few 1 labels in the data:

```
clf_df[['label']].sort_index().head(10)
```

This outputs the following:

	label
Anzelma	0
Babet	0
Bahorel	1
Bamatabois	0
BaronessT	0
Blacheville	0
Bossuet	1
Boulatruelle	0
Brevet	0
Brujon	0

Figure 10.13 – Les Miserables feature engineering: label spot check

Perfect. We have nodes, and they each have a label. A label of **1** means that they are a member of Friends of the ABC, and a label of **0** means that they are not. With that, our training data is ready for use.

Selecting a model

For this exercise, my goal is to simply show you how network data may be useful in ML, not to go into great detail about ML. There are many, many, many thick books on the subject. This is a book about how NLP and networks can be used together to understand the hidden strings that exist around us and the influence that they have on us. So, I am going to speed past the discussion on how different models work. For this exercise, we are going to use one very useful and powerful model that often works well enough. This model is called **Random Forest**.

Random Forest can take both numeric and categorical data as input. Our chosen features should work very well for this exercise. Random Forest is also easy to set up and experiment with, and it's also very easy to learn what the model found most useful for predictions.

Other models would work. I attempted to use **k-nearest neighbors** and had nearly the same level of success, and I'm sure that **Logistic regression** would have also worked well after some additional preprocessing. **XGBoost** and **SVMs** would have also worked. Some of you might also be tempted to use an NN. Please feel free. I chose to not use an NN, as the setup is more difficult and the training time is typically longer, for an occasional tiny boost to accuracy that may also just be a fluke. Experiment with models! It's a good way to learn, even when you are learning what *not* to do.

Preparing the data

We should do a few more data checks. Most importantly, let's check the balance between classes in the training data:

1. Start with the following code:

    ```
    clf_df['label'].value_counts()
    ...
    0    66
    1    11
    Name: label, dtype: int64
    ```

    The data is imbalanced, but not too badly.

2. Let's get this in percentage form, just to make this a little easier to understand:

    ```
    clf_df['label'].value_counts(normalize=True)
    ...
    0    0.857143
    ```

```
1     0.142857
Name: label, dtype: float64
```

It looks like we have about an 86/14 balance between the classes. Not awful. Keep this in mind, because the model should be able to predict with about 86% accuracy just based on the imbalance alone. It won't be an impressive model at all if it only hits 86%.

3. Next, we need to cut up our data for our model. We will use the features as our X data, and the answers as our y data. As the label was added last, this will be simple:

```
X_cols = clf_df.columns[0:-1]
X = clf_df[X_cols]
y = clf_df['label'].values
```

X_cols is every column except for the last one, which is the label. X is a DataFrame containing only X_cols fields, and y is an array of our answers. Don't take my word for it. Let's do a spot check.

4. Run this code:

```
X.head()
```

This will show a DataFrame.

5. Scroll all the way to the right on the DataFrame. If you don't see the **label** column, we are good to go:

```
y[0:5]
```

This will show the first five labels in y. This is an array. We are all set.

Finally, we need to split the data into training data and test data. The training data will be used to train the model, and the test data is completely unknown to the model. We do not care about the model accuracy on the training data. Remember that. We do not care about the model accuracy or any performance metrics about the training data. We only care about how the model does on unseen data. That will tell us how well it generalizes and how well it will work in the wild. Yes, I understand that this model will not be useful in the wild, but that is the idea.

6. We will split the data using the scikit-learn train_test_split function:

```
from sklearn.model_selection import train_test_split
X_train, X_test, y_train, y_test = train_test_split(X, y,
random_state=1337, test_size=0.4)
```

Since we have so little training data, and since there are so few members of Friends of the ABC, I set test_size to 0.4, which is twice as high as the default. If there were less imbalance, I would have reduced this to 0.3 or 0.2. If I really wanted the model to have as much training data as possible and I was comfortable that it would do well enough, I might even experiment with 0.1. But for this exercise, I went with 0.4. That's my reasoning.

This function does a 60/40 split of the data, putting 60% of the data into X_train and y_train, and the other 40% into X_test and y_test. This sets aside 40% of the data as unseen data that the model will not be aware of. If the model does well against this 40% unseen data, then it's a decent model.

We are now ready to train our model and see how well it does!

Training and validating the model

Model training gets the most attention when people talk about ML but it is usually the easiest step, once the data has been collected and prepared. A lot of time and energy can and should be spent on optimizing your models, via **hyperparameter tuning**. Whichever model you are interested in learning about and using, do some research on how to tune the model, and any additional steps required for data preparation.

With this simple network, the default Random Forest model was already optimal. I ran through several checks, and the default model did well enough. Here's the code:

```
from sklearn.ensemble import RandomForestClassifier
clf = RandomForestClassifier(random_state=1337, n_jobs=-1, n_
estimators=100)
clf.fit(X_train, y_train)
train_acc = clf.score(X_train, y_train)
test_acc = clf.score(X_test, y_test)
print('train_acc: {} - test_acc: {}'.format(train_acc, test_
acc))
```

We are using a Random Forest classifier, so we first need to import the model from the sklearn.ensemble module. Random Forest uses an ensemble of decision trees to make its predictions. Each ensemble is trained on different features from the training data, and then a final prediction is made.

Set random_state to whatever number you like. I like 1337, as an old hacker joke. It's *1337*, *leet*, *elite*. Setting n_jobs to -1 ensures that all CPUs will be used in training the model. Setting n_estimators to 100 will allow for 100 ensembles of decision trees. The number of estimators can be experimented with. Increasing it can be helpful, but in this case, it was not.

Finally, I collect and print the training accuracy and test accuracy. How are our scores?

```
train_acc: 1.0 - test_acc: 0.9354838709677419
```

Not bad results on the test set. This set is unseen data, so we want it to be high. As mentioned before, due to class imbalance, the model should at least get 86% accuracy simply due to 86% of the labels being in the majority class. 93.5% is not bad. However, you should be aware of **underfitting** and **overfitting**. If both your train and test accuracy are very low, the model is likely underfitting the data, and you likely need more data. However, if the training accuracy is much higher than the test set accuracy, this can be a sign of overfitting, and this model appears to be overfitting. However, with as little data as we have, and for the sake of this experiment, good enough is good enough today.

It is important that you know that model accuracy is never enough. It doesn't tell you nearly enough about how the model is doing, especially how it is doing with the minority class. We should take a look at the confusion matrix and classification report to learn more. To use both of these, we should first place the predictions for X_test into a variable:

```
predictions = clf.predict(X_test)
predictions

...

array([0, 0, 0, 0, 0, 1, 0, 0, 0, 0, 0, 0, 0, 0, 0, 0, 0, 0, 0,
0, 0, 0,1, 0, 0, 1, 0, 0, 1, 0, 1])
```

Great. We have an array of predictions. Next, let's import the `confusion_matrix` and `classification_report` functions:

```
from sklearn.metrics import confusion_matrix, classification_
report, plot_confusion_matrix
```

We can use both by feeding both of them the X_test data as well as the predictions made against X_test. First, let's look at the confusion matrix:

```
confusion_matrix(y_test, predictions)

...

array([[26,   2],
               [ 0,   3]], dtype=int64)
```

If this isn't clear enough, we can also visualize it:

```
plot_confusion_matrix(clf, X_test, y_test)
```

This produces the following matrix.

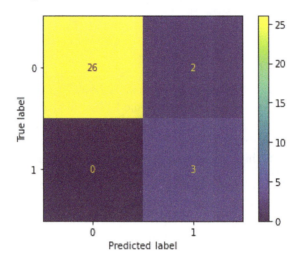

Figure 10.14 – Model confusion matrix

The confusion matrix shows how well a model predicts by classes. The figure depicts this well. The *y-axis* shows the true label of either **0** or **1**, and the *x-axis* shows the predicted label of either **0** or **1**. I can see that 26 of the characters were correctly predicted to not be members of Friends of the ABC (revolutionaries). Our model correctly predicted three of the members of Friends of the ABC, but it also predicted that two of the non-members were members. We should look into that! Sometimes the misses can help us find problems in the data, or they'll shine a light on an interesting insight.

I also find it extremely helpful to take a look at the classification report:

```
report = classification_report(y_test, predictions)
print(report)
```

We get the following output:

	precision	recall	f1-score	support
0	1.00	0.93	0.96	28
1	0.60	1.00	0.75	3
accuracy			0.94	31
macro avg	0.80	0.96	0.86	31
weighted avg	0.96	0.94	0.94	31

Figure 10.15 – Model classification report

This report clearly shows us that the model does very well at predicting non-members of Friends of the ABC, but it does less well at predicting the revolutionaries. Why is that? What is it getting tripped up by? Looking at the networks, the model should have been able to learn that there is a clear difference in the structure of different groups of people, especially comparing the community of Friends of the ABC against everyone else. What gives?! Let's build a simple DataFrame to check:

```
check_df = X_test.copy()
check_df['label'] = y_test
check_df['prediction'] = predictions
check_df = check_df[['label', 'prediction']]
check_df.head()
```

We get the following DataFrame:

	label	prediction
Cochepaille	0	0
Fauchelevent	0	0
Montparnasse	0	0
MlleGillenormand	0	0
Valjean	0	0

Figure 10.16 – DataFrame for prediction checks

Now let's create a mask to look up all rows where the label does not match the prediction:

```
mask = check_df['label'] != check_df['prediction']
check_df[mask]
```

That gives us *Figure 10.17*.

	label	prediction
Joly	0	1
MmeHucheloup	0	1

Figure 10.17 – DataFrame of missed predictions

OK, now we can see what the model got wrong, but why? To save you some time, I looked into both of these characters. Joly actually *is* a member of Friends of the ABC, and Madame Hucheloup runs a cafe where members of Friends of the ABC regularly meet. She was the proprietress of the Corinthe Tavern, the meeting place and last defense of the members! Because she was connected with members of the group, the model predicted that she also was one of them.

To be fair, I bet a human might have made the same judgment about Madame Hucheloup! To me, this is a beautiful misclassification!

The next steps would be to definitely give Joly a correct label and retrain the model. I would leave Madame Hucheloup as is, as she is not a member, but if I were counter-insurgency, I would keep an eye on her.

In short, the model did very well, in my opinion, and it did so entirely using graph data.

Model insights

To me, the model insights are more exciting than building and using the models for prediction. I enjoy learning about the world around me, and ML models (and networks) allow me to understand the world in a way that my eyes do not. We cannot see all of the lines that connect us as people, and we cannot easily understand influence based on how the people around us are strategically placed in the social networks that exist in real life. These models can help with that! Networks can provide the structure to extract contextual awareness of information flow and influence. ML can tell us which of these pieces of information is most useful in understanding something. Sometimes, ML can cut right through the noise and get right to the signals that affect our lives.

With the model that we just built, one insight is that the book *Les Miserables* has different characters by type in different network structures. Revolutionaries are close to each other and tightly connected. Students are also densely connected, and I'm surprised and pleased that the model did not false out on a lot of students. Other characters in the book have very few connections, and their neighbors are sparsely connected. I think it's beautiful that the author put so much work into defining the social network that exists in this story. It can give a new appreciation for the creation of the story.

But what features were most important to the model that helped it make its predictions so well? Let's take a look. Random Forest makes this very easy for us!

You can get to the **feature importances** very easily by doing this:

```
clf.feature_importances_
```

But this data is not very useful in this format. It is much more useful if you put the feature importances into a DataFrame, as they can then be sorted and visualized:

```
importance_df = pd.DataFrame(clf.feature_importances_, index=X_
test.columns)
importance_df.columns = ['value']
importance_df.sort_values('value', ascending=False,
inplace=True)
importance_df.head()
```

We get this DataFrame:

	value
adj_bossuet	0.107379
adj_enjolras	0.107083
adj_combeferre	0.090284
adj_bahorel	0.089942
adj_feuilly	0.086459
adj_joly	0.082105
adj_grantaire	0.080481
triangles	0.070665
adj_courfeyrac	0.061465
adj_prouvaire	0.052652

Figure 10.18 – DataFrame for feature importances

These are the importances in numerical format. This shows the 10 features that the model found most useful in making predictions. Notably, a character's connection to Bossuet and Enjolras was a good indicator of whether a character was a revolutionary or not. Out of the network features, triangles was the only one to make the top 10. The rest of the important features came from the adjacency matrix. Let's visualize this as a bar chart so that we can see more, as well as the level of importance:

```
import matplotlib.pyplot as plt
importance_df[0:20].plot.barh(figsize=(10,8)).invert_yaxis()
```

We get the following plot:

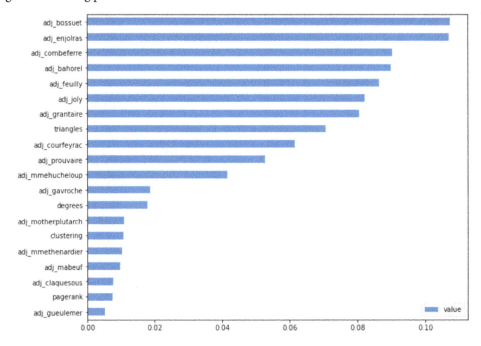

Figure 10.19 – Horizontal bar chart of feature importances

Much better. This is much easier to look at, and it shows exactly how useful the model found each individual feature.

As a side note, you can use feature importances to aggressively identify features that you can cut out of your training data, making for leaner models. I often create a baseline Random Forest model to assist in aggressive feature selection. Aggressive feature selection is what I call it when I ruthlessly cut data that I do not need before training models. For this model, I did not do aggressive feature selection. I used all of the data that I pulled together.

Other use cases

While this may be interesting, most of us don't hunt revolutionaries as part of our day jobs. So, what good is this? Well, there are lots of uses for doing predictions against networks. Lately, graph ML has gotten a lot of interest, but most articles and books tend to showcase models built by other people (not how to build them from scratch), or use NNs. This is fine, but it's complicated, and not always practical.

This approach that I showed is lightweight and practical. If you have network data, you could do something similar.

But what other use cases are there? For me, the ones I'm most interested in are bot detection, and the detection of artificial amplification. How would we do that? For bot detection, you may want to look at features such as the age of the account in days, the number of posts made over time (real people tend to slowly learn how to use a social network before becoming active), and so on. For artificial amplification, you might look at how many followers an account picks up for each tweet that they make. For instance, if an account came online a week ago, made 2 posts, and picked up 20 million followers, what caused that kind of growth? Organic growth is much slower. Perhaps they brought followers from another social network. Perhaps their account was pushed by hundreds of blogs.

What other use cases can you think of? Be creative. You know what networks are now, and you know how to build them and work with them. What would you like to predict or better understand?

Summary

We did it! We made it to the end of another chapter. I truly hope you found this chapter especially interesting because there are so few sources that explain how to do this from scratch. One of the reasons I decided to write this book is because I was hoping that ideas like this would take off. So, I hope this chapter grabbed your attention and sparked your creativity.

In this chapter, we transformed an actual network into training data that we could use for machine learning. This was a simplified example, but the steps will work for any network. In the end, we created a model that was able to identify members of Friends of the ABC revolutionary group, though it was a very simple model and not suitable for anything real-world.

The next chapter is going to be very similar to this one, but we will be using unsupervised ML to identify nodes that are similar to other nodes. Very likely, unsupervised ML will also identify members of Friends of the ABC, but it will likely also bring out other interesting insights.

11

Unsupervised Machine Learning on Network Data

Welcome to another exciting chapter exploring network science and data science together. In the last chapter, we used supervised ML to train a model that was able to detect the revolutionaries from the book *Les Miserables*, using graph features alone. In this chapter, we are going to explore unsupervised ML and how it can also be useful in graph analysis as well as node classification with supervised ML.

The order these two chapters have been written in was intentional. I wanted you to learn how to create your own training data using graphs rather than being reliant on embeddings from unsupervised ML. The reason for this is important: when you rely on embeddings, you lose the ability to interpret why ML models have been classified the way that they have. You lose interpretability and explainability. The classifier essentially works as a black box, no matter which model you use. I wanted to show you the interpretable and explainable approach first.

In this chapter, we will be using a Python library called **Karate Club**. The library is excellent for use in both community detection and the creation of graph embeddings using graph ML. However, there is no way to gain insights into what exactly the model found useful when using this approach. So, I saved it for last. It can still be very effective if you don't mind the loss of interpretability.

This is going to be a fun chapter, as we will be pulling so many things from this book together. We will create a graph, create training data, do community detection, create graph embeddings, do some network visualization, and even do some node classification using supervised ML. If you started the book by reading this chapter, this will all probably look like magic. If you have been following along since *Chapter 1*, this should all make sense and be easy to understand.

Technical requirements

In this chapter, we will be using the Python libraries NetworkX, pandas, scikit-learn, and Karate Club. Other than Karate Club, these libraries should be installed by now, so they should be ready for your use. The steps for installing Karate Club are included in this chapter. If other libraries are not installed, you can install Python libraries with the following:

```
pip install <library name>
```

For instance, to install NetworkX, you would do this:

```
pip install networkx
```

In *Chapter 4*, we also introduced the `draw_graph()` function, which uses both NetworkX and `scikit-network`. You will need that code anytime that we do network visualization. Keep it handy!

All the code is available from the GitHub repo: `https://github.com/PacktPublishing/Network-Science-with-Python`.

What is unsupervised ML?

In books and courses about ML, it is often explained that there are three different kinds: supervised learning, unsupervised learning, and reinforcement learning. Sometimes, combinations will be explained, such as semi-supervised learning. With supervised learning, we provide data (X) and an answer (y), and the model learns to make predictions. With unsupervised learning, we provide data (X), but no answer (y) is given. The goal is for the model to learn to identify patterns and characteristics of the data by itself, and then we use those patterns and characteristics for something else. For instance, we can use unsupervised ML to automatically learn the characteristics of a graph and convert those characteristics into embeddings that we can use in supervised ML prediction tasks. In this situation, an unsupervised ML algorithm is given a graph (G), and it generates embeddings that will serve as the training data (X) that will be used to be able to predict answers.

In short, the goal of unsupervised ML is to identify patterns in data. Often, we call these patterns clusters, but this is not limited to clustering. Creating embeddings is not clustering. However, with embeddings, a complex network has been reduced to a few numeric features that ML will be better able to use.

In this chapter, you'll see firsthand what that actually looks like, as well as the pros and cons of this approach. This is not all positive. There are some less-than-desirable side effects to using embeddings as training data.

Introducing Karate Club

I'm going to showcase a Python library that we have touched on previously in this book: Karate Club. I mentioned it briefly in previous chapters, but now we are going to actually use it. I purposefully held off on going into detail before now, because I wanted to teach core approaches to working with networks before showing seemingly easy approaches to extracting communities and embeddings from networks using ML. This is because there are some undesirable side effects to using network embeddings rather than metrics extracted from a network. I will get into that in a bit. For now, I want to introduce this awesome, performant, and reliable Python library.

Karate Club's documentation (`https://karateclub.readthedocs.io/en/latest/`) gives a clear and concise explanation of what the library does:

> *Karate Club is an unsupervised machine learning extension library for NetworkX. It builds on other open source linear algebra, machine learning, and graph signal processing libraries such as NumPy, SciPy, Gensim, PyGSP, and Scikit-learn. Karate Club consists of state-of-the-art methods to do unsupervised learning on graph-structured data. To put it simply, it is a Swiss Army knife for small-scale graph mining research.*

Two things should stand out from this paragraph: *unsupervised machine learning* and *graph*. You can think of Karate Club simply as unsupervised learning for graphs. The outputs of Karate Club can then be used with other libraries for actual prediction.

There are so many cool approaches to unsupervised learning stacked into Karate Club that it is a real thrill to learn about them. You can learn about them at `https://karateclub.readthedocs.io/en/latest/modules/root.html`. The thing that I love the most is that the documentation links to the original research papers that were written about the algorithms. This allows you to really get to know the processes behind unsupervised ML models. To pick out models to use for this chapter, I read seven research papers, and I loved every moment of it.

Another nice thing about this library is that the outputs are standardized across models. The embeddings generated by one model will be like the embeddings generated by another model. This means that you can easily experiment with different approaches for embeddings, and see how they affect models used for classification. We will do exactly that in this chapter.

Finally, I've never seen community detection as simple as I have with Karate Club. Using NetworkX or other libraries for Louvain community detection can take a little work to set up. Using **Scalable Community Detection (SCD)** from Karate Club, you can go from a graph to identified communities in very few lines of code. It's so clean.

If you want to learn more about Karate Club and graph machine learning, I recommend the book *Graph Machine Learning*. You can pick up a copy at `https://www.amazon.com/Graph-Machine-Learning-techniques-algorithms/dp/1800204493/`. The book goes into much greater detail on Karate Club's capabilities than this chapter will be able to. It is also a good follow-up book to read after this book, as this book teaches the fundamentals of interacting with networks using Python, and *Graph Machine Learning* takes this knowledge a step further.

Network science options

It is important to know that you do not *need* to use ML to work with graphs. ML can just be useful. There is also a blurry line between what is and isn't ML. For instance, I would consider any form of community detection to be unsupervised ML, as these algorithms are capable of automatically identifying communities that exist in a network. By that definition, we could consider some of the approaches offered by NetworkX unsupervised ML, but they are not given the same level of attention in the data science community, because they are not explicitly called graph ML. There is a level of hype to be aware of.

I am saying this because I want you to keep in mind that there are approaches that you have already learned that can eliminate the need to use what is advertised as graph ML. For instance, you can use Louvain to identify communities, or even just connected components. You can use PageRank to identify hubs – you don't need embeddings for that. You can use `k_corona(0)` to identify isolates – you don't need ML at all for that. You can chain together several graph features into training data, like we did in the last chapter. You don't *need* to use Karate Club to create embeddings, and you *shouldn't* use Karate Club embeddings if you are interested in any kind of model interpretability.

Remember what you have learned in this book for interrogating and dissecting networks. Use what is in this chapter as a shortcut or if the science behind what you are doing is already figured out. Embeddings can be a nice shortcut, but any model using these embeddings will become a non-interpretable black box.

My recommendation: use network science approaches (in NetworkX) rather than Karate Club when possible, but be aware of Karate Club and that it can be useful. This suggestion isn't due to any disdain for Karate Club. It's because I find the insights I can extract from models to be illuminating, and almost nothing is worth losing those insights to me. For instance, what characteristics allow a model to predict bots and artificial amplification?

Loss of interpretability means that you won't be able to understand your model's behavior. This is never a good thing. This is not a dig at approaches that decompose graphs into embeddings, or the research around those approaches; it is just worth knowing that certain approaches can lead to a total loss of interpretability of model behavior.

Uses of unsupervised ML on network data

If you take a look at the Karate Club website, you will probably notice that the two approaches to unsupervised ML fall into two categories: identifying communities or creating embeddings. Unsupervised ML can be useful for creating embeddings not just for nodes, but also for edges or for whole graphs.

Community detection

Community detection is the easiest to understand. The goal of using a community detection algorithm is to identify the communities of nodes that exist in a network. You can think of communities as clusters or clumps of nodes that interact with each other in some way. In social network analysis, this is called community detection, because it is literally about identifying communities in a social network. However, community detection can be useful outside of social network analysis involving people. Maybe it helps to think of a graph as just a social network of things that somehow interact. Websites interact. Countries and cities interact. People interact. There are communities of countries and cities that interact (allies and enemies). There are communities of websites that interact. There are communities of people that interact. It's just about identifying groups of things that interact.

We discuss community detection in *Chapter 9* of this book. If you haven't yet read that, I encourage you to go back to that chapter to learn more about it.

Here is an example community to refresh your memory:

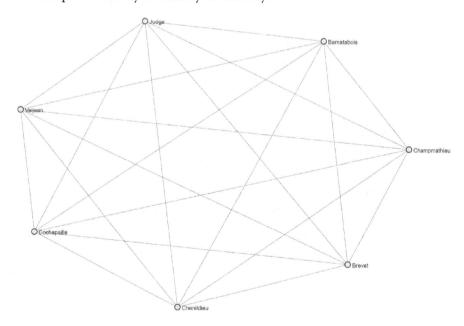

Figure 11.1 – Community from Les Miserables

Looking at this community, we can see that it is tightly knit. Each member is connected with every other member of the community. Other communities are more sparsely connected.

Graph embeddings

I like to think about graph embeddings as a translation of a complex network into a data format that mathematical models will be better able to use. For instance, if you use a graph edge list or a NetworkX graph (G) with Random Forest, nothing is going to happen. The model will have no way of using the input data for anything. In order to make use of these models, we need to deconstruct graphs into a more usable format. In the previous chapter on supervised machine learning, we converted a graph into training data in this format:

	degrees	clustering	triangles	betw_cent	close_cent	pagerank
Napoleon	1	0.000000	0	0.000000	0.301587	0.005584
Myriel	10	0.066667	3	0.176842	0.429379	0.042803
MlleBaptistine	3	1.000000	3	0.000000	0.413043	0.010279
MmeMagloire	3	1.000000	3	0.000000	0.413043	0.010279
CountessDeLo	1	0.000000	0	0.000000	0.301587	0.005584
Geborand	1	0.000000	0	0.000000	0.301587	0.005584
Champtercier	1	0.000000	0	0.000000	0.301587	0.005584
Cravatte	1	0.000000	0	0.000000	0.301587	0.005584
Count	1	0.000000	0	0.000000	0.301587	0.005584
OldMan	1	0.000000	0	0.000000	0.301587	0.005584

Figure 11.2 – Hand-crafted graph training data

We also included a **label**, which is the answer that an ML model will learn from. After this, we tacked on the adjacency matrix for each node, so that the classification model would also learn from network connections.

As you can see, it's easy for us to know what the features are in this training data. First, we have a node's degrees, then its clustering, the number of triangles, its betweenness and closeness centrality, and finally, its PageRank score.

With embeddings, all of the information in a graph is deconstructed into a series of embeddings. If you read the article behind the model, you can get an understanding of what is happening in the process, but by the time the embeddings are created, it's really not super interpretable. This is what embeddings look like:

```
eb_df = pd.DataFrame(embeddings, index=nodes)
eb_df['label'] = clf_df['label']
eb_df.head()
```

	0	1	2	3	4	5	6	7	8	9	...	1741	1742	1743
Napoleon	0.664985	0.471134	0.136111	-0.016215	0.154984	0.469122	0.601893	0.404308	0.049391	-0.137721	...	0.006781	0.006675	0.006496	C
Myriel	0.910804	0.723366	0.656674	0.610370	0.466355	0.628329	0.440952	0.388751	0.242645	0.030653	...	0.018148	0.017863	0.017386	C
MlleBaptistine	0.786000	0.345126	0.209002	0.136510	-0.156203	0.299175	-0.070888	-0.104387	-0.352101	-0.747891	...	0.036940	0.036360	0.035390	C
MmeMagloire	0.786000	0.345126	0.209002	0.136510	-0.156203	0.299175	-0.070888	-0.104387	-0.352101	-0.747891	...	0.036940	0.036360	0.035390	C
CountessDeLo	0.664985	0.471134	0.136111	-0.016215	0.154984	0.469122	0.601893	0.404308	0.049391	-0.137721	...	0.006781	0.006675	0.006496	C

5 rows × 1751 columns

Figure 11.3 – Unsupervised ML graph embeddings

Sweet, we've converted a graph into 1,751 columns of… what?

Still, these embeddings are useful and can be fed directly to supervised ML models for prediction, and the predictions can be quite useful, even if the model and data are not very interpretable.

But what can these embeddings be used for? They're just a whole bunch of columns of numeric data with no description. How can that be useful? Well, there are two downstream uses, one involving more unsupervised ML for clustering, and another using supervised ML for classification.

Clustering

In **clustering**, your goal is to identify clusters, clumps, or groups of things that look or behave similarly. With both the hand-crafted training data and the Karate Club-generated embeddings, clustering is possible. Both of these datasets can be fed to a clustering algorithm (such as K-means) to identify similar nodes, for example. There are implications to using any model, though, so spend time learning about the models you are interested in using. For instance, to use K-means, you have to specify the number of clusters that you expect to exist in the data, and that is practically never known.

Getting back to the **Keep it Simple (KISS)** approach, though, stacking math upon math upon math just to get a result that you could have gotten simply by knowing the right network approach is extremely wasteful. If you just wanted to identify clusters, you could have used a community detection algorithm, k_corona, looked at the connected components, or sorted nodes by PageRank. If you are using ML, you should first ask yourself whether there is a network-based approach to this that eliminates the need for ML.

Classification

With classification, your goal is to predict something. In a social network, you might want to predict who will eventually become friends, who might like to become friends, who might click on an advertisement, or who might want to buy a product. If you can make these predictions, you can automate recommendations and ad placement. Or, you might want to identify fraud, artificial amplification, or abuse. If you can make these predictions, you can automatically quarantine what looks like bad behavior, and automate the fielding of responding to these kinds of cases.

Classification usually gets the most attention in ML. It deserves the glory. With classification, we can prevent spam from ruining our inboxes and productivity, we can automatically translate text from one language into another, and we can prevent malware from ruining our infrastructure and allowing criminals to take advantage of us. Classification can literally make the world a better and safer place when used well and responsibly.

In the previous chapter, we invented a fun game called "Spot the Revolutionary." The same game could be played in real life with different purposes. You could automatically spot the influencer, spot the fraudulent behavior, spot the malware, or spot the cyber attack. Not all classifiers are deadly serious. Some classifiers help us learn more about the world around us. For instance, if you are using hand-crafted training data rather than embeddings, you could train a model to predict bot-like behavior, and then you could learn what features the model found most useful in identifying bots. For instance, maybe the fact that a bot account was created two days ago, has done zero tweets, has done 2,000 retweets, and already has 15,000 followers could have something to do with it. A model trained on embeddings might tell you that embedding number 72 was useful, which means nothing.

Alright, enough talk. Let's get to coding and see all of this in action. For the rest of this chapter, we will be using Karate Club approaches.

Constructing a graph

Before we can do anything, we need a graph to play with. As with the last chapter, we will make use of the NetworkX *Les Miserables* graph, for familiarity.

First, we'll create the graph and remove the additional fields that we don't need:

```
import networkx as nx
import pandas as pd


G = nx.les_miserables_graph()
df = nx.to_pandas_edgelist(G)[['source', 'target']] # dropping
'weight'
G = nx.from_pandas_edgelist(df)
```

```
G_named = G.copy()
G = nx.convert_node_labels_to_integers(G, first_label=0,
ordering='default', label_attribute=None)
nodes = G_named.nodes
```

If you look closely, I've included two lines of code that create a G_named graph as a copy of G, and have converted node labels on graph G to numbers for use in Karate Club a bit later in this chapter. This is a required step for working with Karate Club.

Let's visualize graph G for a sanity check:

```
draw_graph(G, node_size=4, edge_width=0.2)
```

This produces the following graph. We are not including node labels, so it will just be dots and lines (nodes and edges).

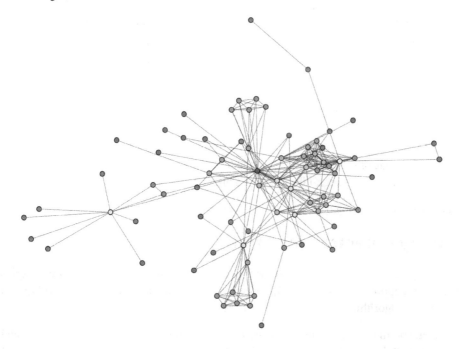

Figure 11.4 – Les Miserables network

This looks as expected. Each node has a label, but we are not showing them.

I have also created some training data with labels. The data is included in the /data section of the GitHub repo accompanying this book:

```
train_data = 'data/clf_df.csv'

clf_df = pd.read_csv(train_data).set_index('index')
```

The process of creating the training data is a bit involved and was explained in the previous chapter, so please use those steps to learn how to do this manually. For this chapter, you can just use the CSV file to save time. Let's check that the data looks correct:

```
clf_df.head()
```

	degrees	clustering	triangles	betw_cent	close_cent	pagerank	label
Napoleon	1	0.000000	0	0.000000	0.301587	0.005584	0
Myriel	10	0.066667	3	0.176842	0.429379	0.042803	0
MlleBaptistine	3	1.000000	3	0.000000	0.413043	0.010279	0
MmeMagloire	3	1.000000	3	0.000000	0.413043	0.010279	0
CountessDeLo	1	0.000000	0	0.000000	0.301587	0.005584	0

Figure 11.5 – Hand-crafted training data

In this chapter, only one of the models will use the hand-crafted training data as input, but we will use the labels with our embeddings. I'll show how a bit later.

With the graph and the training data, we are set to continue.

Community detection in action

With community detection, our obvious goal is to identify the communities that exist in a network. I explained various approaches in *Chapter 9, Community Detection*. In this chapter, we will make use of two Karate Club algorithms: SCD and EgoNetSplitter.

For this chapter, and in general, I tend to gravitate toward models that can scale well. If a model or algorithm is only useful on a tiny network, I'll avoid it. Real networks are large, sparse, and complicated. I don't think I've ever seen something that doesn't scale well that is actually better than algorithms that do. This is especially true in community detection. The best algorithms do scale well. My least favorite do not scale well at all.

SCD

The first community detection algorithm I want to showcase is SCD. You can find the documentation and journal article about the model at `https://karateclub.readthedocs.io/en/latest/modules/root.html#karateclub.community_detection.non_overlapping.scd.SCD`.

This model claims to be much faster than the most accurate state-of-the-art community detection solutions while retaining or even exceeding their quality. It also claims to be able to handle graphs with billions of edges, which means that it can be useful with real-world networks. It claims to perform better than Louvain, the fastest community detection algorithm.

Those are some bold claims. Louvain is extremely useful for community detection for a few reasons. First, it is very fast and useful on large networks. Second, the Python implementation is simple to work with. So, we already know that Louvain is fast and easy to work with. How much better is this? Let's try it out:

1. First, make sure you have Karate Club installed on your computer. You can do so with a simple `pip install karateclub`.

2. Now, let's use the model. First, start with the imports. You need these two:

   ```
   from karateclub.community_detection.non_overlapping.scd
   import SCD

   import numpy as np
   ```

3. Now that we have those, getting the communities for our graph is as simple as 1, 2, 3:

   ```
   model = SCD()
   model.fit(G)
   clusters = model.get_memberships()
   ```

   We first instantiate SCD, then we fit the graph to SCD, and then we get the cluster memberships for each node. Karate Club models are this simple to work with. You need to read the articles to know what is happening under the hood.

4. What do the clusters look like? If we print the `clusters` variable, we should see this:

   ```
   {0: 34,
    1: 14,
    2: 14,
    3: 14,
    4: 33,
    5: 32,
    6: 31,
   ```

```
 7: 30,
 8: 29,
 9: 28,
10: 11,
...
}
```

Node zero is in cluster 34, nodes 1-3 are in cluster 14, node 4 is in cluster 33, and so forth.

5. Next, we shove the clusters into a numpy array so that we can use the data with our named nodes to more easily determine what nodes belong to what clusters:

```
clusters = np.array(list(clusters.values()))
```

The clusters variable will now look like this:

```
array([34, 14, 14, 14, 33, 32, 31, 30, 29, 28, 11, 27,
13, 26, 25, 24,  7,
       15, 15,  4, 15,  9, 11,  6, 23, 35, 11, 11, 11,
11, 11, 36,  9,  1,
        4,  4,  1,  1,  1, 15, 15, 15, 15,
37,  7,  7,  7,  7,  7,  7,  7,
        6, 15, 15, 22, 17, 21, 15,  4, 20, 17,  1,  1,
19, 19,  1,  1,  1,
        1,  1,  1,  2,  2,  1,  0, 18, 16])
```

6. Then, we create a cluster DataFrame:

```
cluster_df = pd.DataFrame({'node':nodes,
'cluster':clusters})
cluster_df.head(10)
```

This gives us the following output:

	node	cluster
0	Napoleon	34
1	Myriel	14
2	MlleBaptistine	14
3	MmeMagloire	14
4	CountessDeLo	33
5	Geborand	32
6	Champtercier	31
7	Cravatte	30
8	Count	29
9	OldMan	28

Figure 11.6 – SCD cluster DataFrame

Great. This is much easier to understand in this format. We now have actual people nodes as well as the community that they belong to.

7. Let's find the largest communities by node membership:

```
title = 'Clusters by Node Count (SCD)'
cluster_df['cluster'].value_counts()[0:10].plot.
barh(title=title).invert_yaxis()
```

This gives us the following:

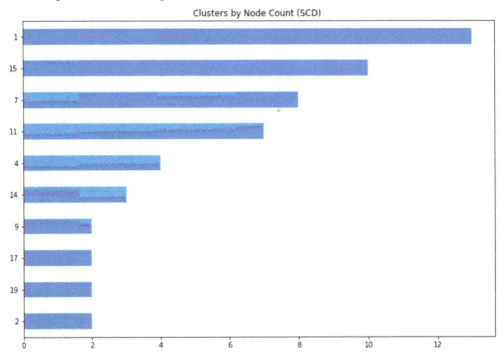

Figure 11.7 – SCD communities by node count

8.   Community 1 is the largest, with 13 members, followed by community 15, which has 10 members. Let's examine both of these:

```
check_cluster = 1
community_nodes = cluster_df[cluster_
df['cluster']==check_cluster]['node'].to_list()
G_comm = G_named.subgraph(community_nodes)
draw_graph(G_comm, show_names=True, node_size=5)
```

This gives us the following:

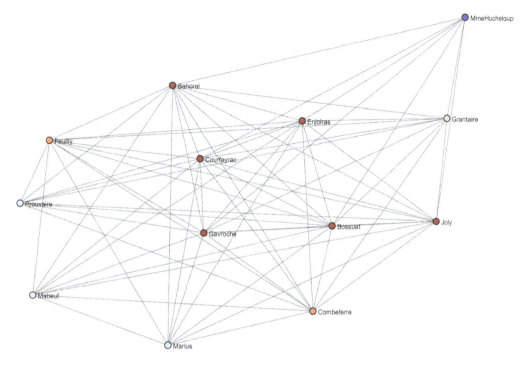

Figure 11.8 – SCD community 1

This is excellent. This is a clear community of highly connected nodes. This is a densely connected community. Not all nodes are equally well connected. Some nodes are more central than others.

9. Let's look at community 15:

```
check_cluster = 15
community_nodes = cluster_df[cluster_
df['cluster']==check_cluster]['node'].to_list()
G_comm = G_named.subgraph(community_nodes)
draw_graph(G_comm, show_names=True, node_size=5)
```

The results follow:

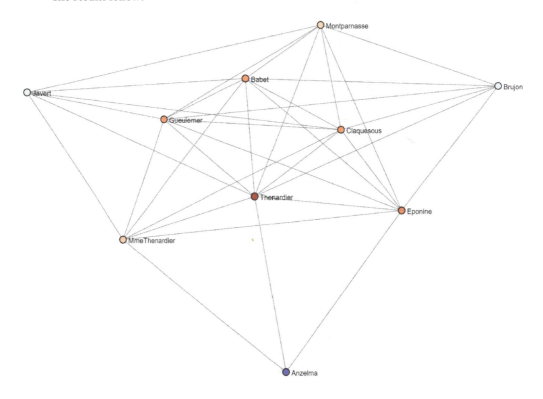

Figure 11.9 – SCD community 15

This is another high-quality community extraction. All nodes are connected to other nodes in the community. Some nodes are more central than others.

10. Let's look at one more community:

```
check_cluster = 7
community_nodes = cluster_df[cluster_
df['cluster']==check_cluster]['node'].to_list()
G_comm = G_named.subgraph(community_nodes)
draw_graph(G_comm, show_names=True, node_size=5)
```

We get *Figure 11.10*.

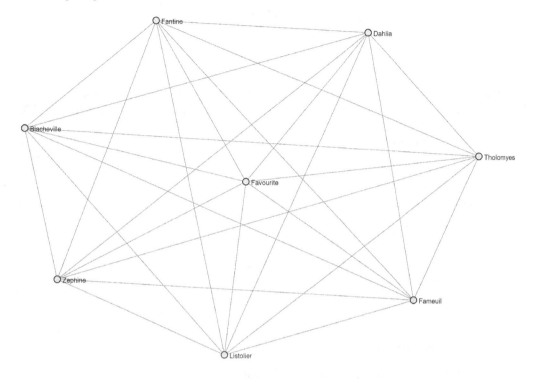

Figure 11.10 – SCD community 7

This is another high-quality community extraction. All nodes in the community are connected. In this case, this is quite a pleasing visualization to look at, as all nodes are equally connected. It is quite symmetric and beautiful.

The *Les Miserables* network is tiny, so naturally, the SCD model was able to train on it essentially instantly.

One thing that I do like about this approach is that the setup is simpler than the approaches I explained in *Chapter 9*. I can go from a graph to communities in no time and with very little code. The fact that this can supposedly scale to networks with billions of edges is incredible, if true. It is fast, clean, and useful.

EgoNetSplitter

The next model we will test for community detection is named EgoNetSplitter. You can learn about it here: `https://karateclub.readthedocs.io/en/latest/modules/root.html#karateclub.community_detection.overlapping.ego_splitter.EgoNetSplitter`.

In Jupyter, if you *Shift + Tab* into the model instantiation code, you can read about it:

> *The tool first creates the ego-nets of nodes. A persona-graph is created which is clustered by the Louvain method. The resulting overlapping cluster memberships are stored as a dictionary.*

So, this model creates ego networks, then uses Louvain for clustering, and then overlapping memberships are stored as a dictionary. It's an interesting approach and different from other approaches, so I thought it'd be neat to test it out and see how it performs. The steps are slightly different from those with SCD:

1. To begin, let's get the model in place:

```
from karateclub.community_detection.overlapping.ego_
splitter import EgoNetSplitter
model = EgoNetSplitter()
model.fit(G)
clusters = model.get_memberships()
clusters = np.array(list(clusters.values()))
clusters = [i[0] for i in clusters] # needed because put
clusters into an array of arrays
```

2. This gets us our clusters. Then, creating our `cluster` DataFrame and doing visualizations follows the same code as with SCD:

```
cluster_df = pd.DataFrame({'node':nodes,
'cluster':clusters})
```

3. Let's check community membership by count:

```
title = 'Clusters by Node Count (EgoNetSplitter)'
cluster_df['cluster'].value_counts()[0:10].plot.
barh(title=title).invert_yaxis()
```

We get the following output:

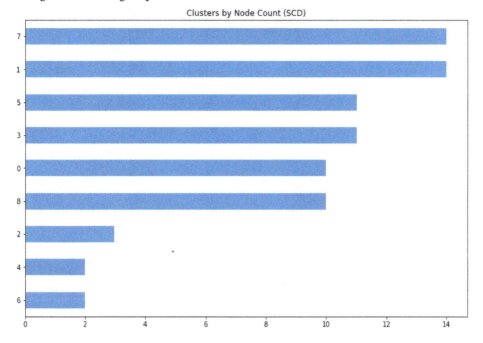

Figure 11.11 – EgoNetSplitter communities by node count

Already, the results look different from SCD. This should be interesting.

4. Let's take a look to see what is different. Clusters 7 and 1 are the largest, so let's take a look at those two:

```
check_cluster = 7
community_nodes = cluster_df[cluster_
df['cluster']==check_cluster]['node'].to_list()
G_comm = G_named.subgraph(community_nodes)
draw_graph(G_comm, show_names=True, node_size=5)
```

This will draw our ego network.

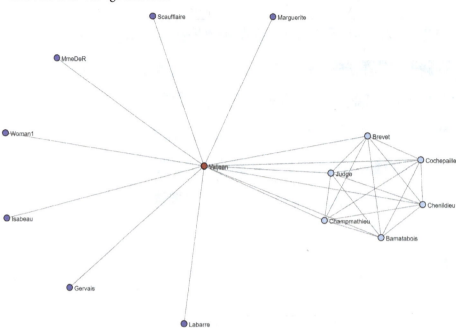

Figure 11.12 – EgoNetSplitter community 7

I don't like that. I don't believe that the nodes on the left should be a part of the same community as the nodes that are connected to the densely connected nodes on the right. Personally, I don't find this to be as useful as SCD's results.

5. Let's take a look at the next most populated cluster:

```
check_cluster = 1
community_nodes = cluster_df[cluster_
df['cluster']==check_cluster]['node'].to_list()
G_comm = G_named.subgraph(community_nodes)
draw_graph(G_comm, show_names=True, node_size=5)
```

Figure 11.13 shows the resulting output.

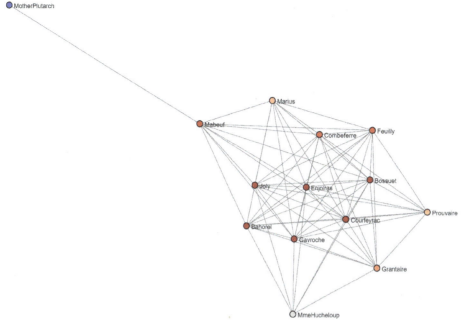

Figure 11.13 – EgoNetSplitter community 1

Once again, we are seeing similar behavior, where one node has been included in the network that really should not be. **MotherPlutarch** might be connected to **Mabeuf**, but she really has nothing to do with the other people in the community.

6. Let's take a final look at the next community:

```
check_cluster = 5
community_nodes = cluster_df[cluster_
df['cluster']==check_cluster]['node'].to_list()
G_comm = G_named.subgraph(community_nodes)
draw_graph(G_comm, show_names=True, node_size=5)
```

The code produces the following output:

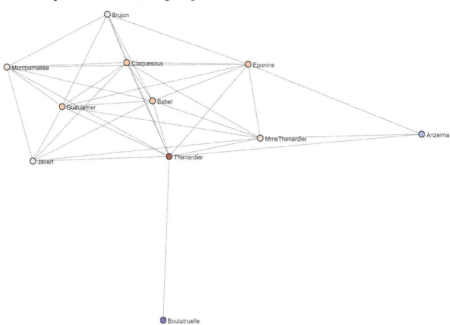

Figure 11.14 – EgoNetSplitter community 5

Again, we see one node connected to one other node, but not connected to the rest of the nodes in the network.

I don't want to say that EgoNetSplitter is inferior to SCD or any other model. I'd say that I personally prefer the outputs of SCD over EgoNetSplitter for community detection. However, it could be argued that it is better to include the few extra nodes as part of the community due to their one connection than it would be to leave them out. It's important to know the difference between the two approaches, as well as the differences in their results.

However, due to the scalability claims of SCD and due to its clean separation of communities, I lean toward SCD for community detection.

Now that we have explored using unsupervised ML for community detection, let's move on to using unsupervised ML for creating graph embeddings.

Graph embeddings in action

Now that we are past the comfort of community detection, we are getting into some weird territory with graph embeddings. The simplest way I think of graph embeddings is just the deconstruction of a complex network into a format more suitable for ML tasks. It's the translation of a complex data structure into a less complex data structure. That's a simple way of thinking about it.

Some unsupervised ML models will create more dimensions (more columns/features) of embeddings than others, as you will see in this section. In this section, we are going to create embeddings, inspect nodes that have similar embeddings, and then use the embeddings with supervised ML to predict "revolutionary or not," like our "Spot the Revolutionary" game from the last chapter.

We're going to quickly run through the use of several different models – this chapter would be hundreds of pages long if I went into great detail about each model. So, to save time, I'll provide the link to the documentation and a simple summary, and we'll just do some simple comparisons. Please know that you should never use ML this blindly. Please read the documentation, read the articles, and know how the models work. I did the legwork, and you should too. Do feel free to just play around with different models to see how they behave. If you are just experimenting and not putting them into production, you aren't going to accidentally cause a rip in the fabric of spacetime by playing with a `scikit-learn` model.

We're going to need this helper function for visualizations of the upcoming embeddings:

```
def draw_clustering(embeddings, nodes, title):

    import plotly.express as px
    from sklearn.decomposition import PCA

    embed_df = pd.DataFrame(embeddings)

    # dim reduction, two features; solely for visualization
    model = PCA(n_components=2)
    X_features = model.fit_transform(embed_df)
    embed_df = pd.DataFrame(X_features)
    embed_df.index = nodes
    embed_df.columns = ['x', 'y']
    fig = px.scatter(embed_df, x='x', y='y', text=embed_
df.index)
    fig.update_traces(textposition='top center')
```

```
    fig.update_layout(height=800, title_text=title, font_
size=11)
    return fig.show()
```

I need to explain a few things. First, this `draw_clustering` function uses `plotly` to create an interactive scatter plot. You can zoom in and out and inspect nodes interactively. You will need to have `plotly` installed, which can be done with `pip install plotly`.

Second, I'm using **Principal Component Analysis (PCA)** to reduce embeddings into two dimensions, just for the sake of visualization. PCA is also unsupervised learning and useful for dimension reduction. I needed to do this so that I could show you that these embedding models behave differently. Reducing embeddings to two dimensions allows me to visualize them on a scatter plot. I do not recommend doing PCA after creating embeddings. I am only using this process for visualization.

FEATHER

The first algorithm we will use is called **FEATHER**, and you can learn about it at `https://karateclub.readthedocs.io/en/latest/modules/root.html#karateclub.node_embedding.attributed.feathernode.FeatherNode`.

In Jupyter, if you *Shift + Tab* into the model instantiation code, you can read about it:

> *An implementation of "FEATHER-N"* `<https://arxiv.org/abs/2005.07959>` *from the CIKM '20 paper "Characteristic Functions on Graphs: Birds of a Feather, from Statistical Descriptors to Parametric Models". The procedure uses characteristic functions of node features with random walk weights to describe node neighborhoods.*

FEATHER claims to create high-quality graph representations, perform transfer learning effectively, and scale well to large networks. It creates node embeddings.

This is actually a very interesting model, as it is able to take both a graph and additional training data for use in creating embeddings. I would love to explore that idea more, to see how well it does with different kinds of training data such as `tf-idf` or topics:

1. For now, let's give it the hand-crafted training data that we used before:

    ```
    from karateclub.node_embedding.attributed.feathernode
    import FeatherNode
    model = FeatherNode()
    model.fit(G, clf_df)
    embeddings = model.get_embedding()
    ```

First, we import the model, then we instantiate it. On the `model.fit` line, notice that we are passing in both G and `clf_df`. The latter is the training data that we created by hand. With every other model, we only pass in G. To me, this is fascinating, as it seems like it'd give the model the ability to learn more about the network based on other contextual data.

2. Let's visualize these embeddings to see how the model is working:

```
title = 'Les Miserables Character Similarity
(FeatherNode)'
draw_clustering(embeddings, nodes, title)
```

We get the following output:

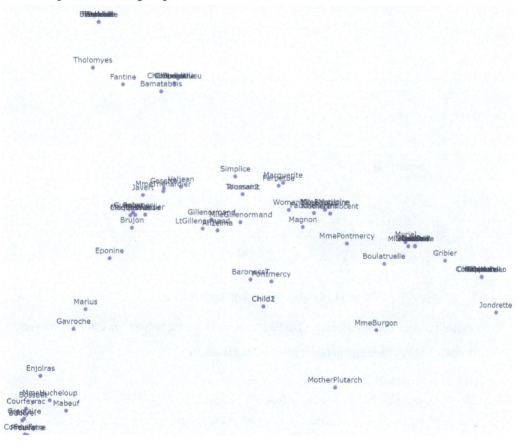

Figure 11.15 – FEATHER embeddings

This is interesting to look at. We can see that there are several nodes that appear together. As this is an interactive visualization, we can inspect any of them. If we zoom in on the bottom-left cluster, we can see this:

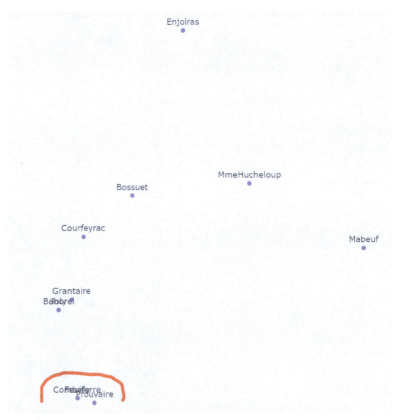

Figure 11.16 – FEATHER embeddings zoomed

It's difficult to read, due to the overlap, but **Feuilly** is shown on the bottom left, close to **Prouvaire**.

3. Let's check both of their ego networks to see what is similar:

```
node = 'Feuilly'
G_ego = nx.ego_graph(G_named, node)
draw_graph(G_ego, show_names=True, node_size=3)
```

That produces *Figure 11.17*:

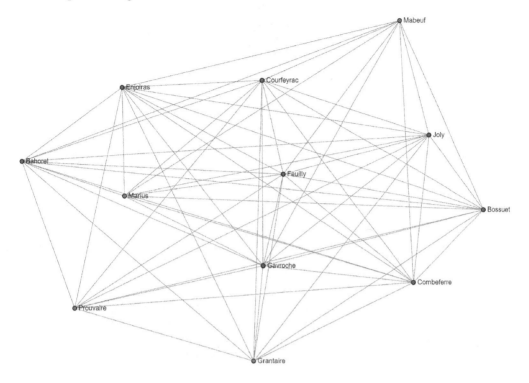

Figure 11.17 – Feuilly ego network

4. Now, let's inspect Prouvaire's ego network:

```
node = 'Prouvaire'
G_ego = nx.ego_graph(G_named, node)
draw_graph(G_ego, show_names=True, node_size=3)
```

This outputs *Figure 11.18*.

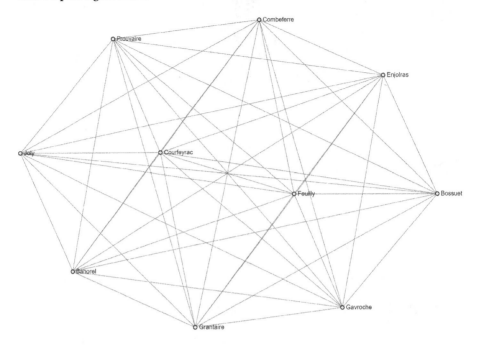

Figure 11.18 – Prouvaire ego network

Nice. The first observation is that they are both part of each other's ego network and also part of each other's community. Second, their nodes are quite connected. On both ego networks, both nodes show as being quite well connected and also part of a densely connected community.

5. Let's take a look at a few other nodes:

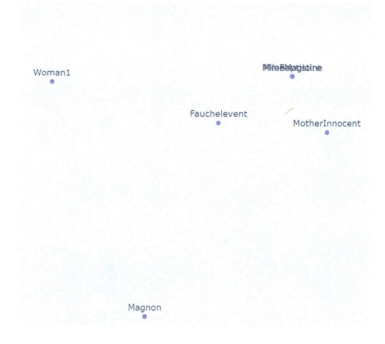

Figure 11.19 – FEATHER embeddings zoomed

6. Let's inspect the ego networks for **MotherInnocent** and **MmeMagloire**. **MotherInnocent** first:

```
node = 'MotherInnocent'
G_ego = nx.ego_graph(G_named, node)
draw_graph(G_ego, show_names=True, node_size=3)
```

Figure 11.20 shows the output.

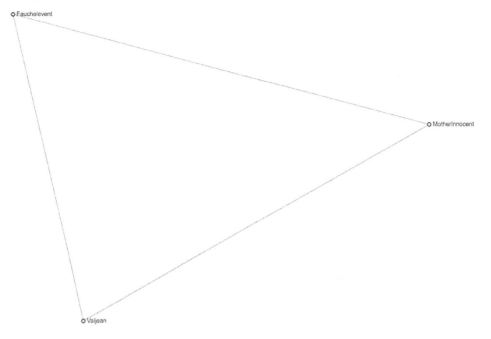

Figure 11.20 – MotherInnocent ego network

And now **MmeMagloire**:

```
node = 'MmeMagloire'
G_ego = nx.ego_graph(G_named, node)
draw_graph(G_ego, show_names=True, node_size=3)
```

Figure 11.21 shows the results.

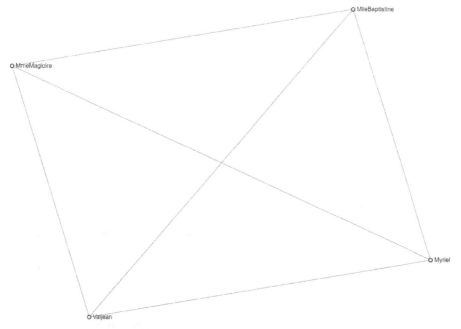

Figure 11.21 – MmeMagloire ego network

MotherInnocent has two edges, and **MmeMagloire** has three. Their ego networks are quite small. These similarities are being picked up by FEATHER and translated into embeddings.

7.   But what do the actual embeddings look like?

```
eb_df = pd.DataFrame(embeddings, index=nodes)
eb_df['label'] = clf_df['label']
eb_df.head(10)
```

This produces the following DataFrame.

	0	1	2	3	4	5	6	7	8	9	...	1741	1742	1743	
Napoleon	0.664985	0.471134	0.136111	-0.016215	0.154984	0.469122	0.601893	0.404308	0.049391	-0.137721	...	0.006781	0.006675	0.006496	0
Myriel	0.910804	0.723366	0.656674	0.610370	0.466355	0.628329	0.440952	0.388751	0.242645	0.030653	...	0.018148	0.017863	0.017386	0
MlleBaptistine	0.786000	0.345126	0.209002	0.136510	-0.156203	0.299175	-0.070888	-0.104387	-0.352101	-0.747891	...	0.036940	0.036360	0.035390	0
MmeMagloire	0.786000	0.345126	0.209002	0.136510	-0.156203	0.299175	-0.070888	-0.104387	-0.352101	-0.747891	...	0.036940	0.036360	0.035390	0
CountessDeLo	0.664985	0.471134	0.136111	-0.016215	0.154984	0.469122	0.601893	0.404308	0.049391	-0.137721	...	0.006781	0.006675	0.006496	0
Geborand	0.664985	0.471134	0.136111	-0.016215	0.154984	0.469122	0.601893	0.404308	0.049391	-0.137721	...	0.006781	0.006675	0.006496	0
Champtercier	0.664985	0.471134	0.136111	-0.016215	0.154984	0.469122	0.601893	0.404308	0.049391	-0.137721	...	0.006781	0.006675	0.006496	0
Cravatte	0.664985	0.471134	0.136111	-0.016215	0.154984	0.469122	0.601893	0.404308	0.049391	-0.137721	...	0.006781	0.006675	0.006496	0
Count	0.664985	0.471134	0.136111	-0.016215	0.154984	0.469122	0.601893	0.404308	0.049391	-0.137721	...	0.006781	0.006675	0.006496	0
OldMan	0.664985	0.471134	0.136111	-0.016215	0.154984	0.469122	0.601893	0.404308	0.049391	-0.137721	...	0.006781	0.006675	0.006496	0

10 rows × 1751 columns

Figure 11.22 – FEATHER embeddings DataFrame

The graph was translated into 1,750 embedding dimensions. In this format, you can think of them as columns or features. A simple network was converted into 1,750 columns, which is quite a lot of data for such a small network. Pay attention to the number of dimensions created by these models as we go through the others after FEATHER.

These embeddings are useful for classification, so let's do just that. I'm going to just throw data at a classification model and hope for the best. This is never a good idea other than for simple experimentation, but that is exactly what we are doing. I encourage you to dig deeper into any of these models that you find interesting.

8. The preceding code already added the `label` field, but we need to create our X and y data for classification:

```
from sklearn.model_selection import train_test_split
X_cols = [c for c in eb_df.columns if c != 'label']
X = eb_df[X_cols]
y = eb_df['label']
X_train, X_test, y_train, y_test = train_test_split(X, y,
random_state=1337, test_size=0.3)
```

X is our data, y is the correct answer. This is our "Spot the Revolutionary" training data. Nodes that are labeled as revolutionaries have a y of 1. The rest have a y of 0.

9. Let's train a Random Forest model, as I want to show you something about interpretability:

```
from sklearn.ensemble import RandomForestClassifier
clf = RandomForestClassifier(random_state=1337)
clf.fit(X_train, y_train)
train_acc = clf.score(X_train, y_train)
test_acc = clf.score(X_test, y_test)
print('train accuracy: {}\ntest accuracy: {}'.
format(train_acc, test_acc))
```

If we run this code, we get these results:

train accuracy: 0.9811320754716981

test accuracy: 1.0

Using the FEATHER embeddings as training data, the model was able to correctly spot the revolutionary 100% of the time on unseen data. This is a tiny network, though, and never, ever, *ever* trust a model that gives 100% accuracy on anything. A model that appears to be hitting 100% accuracy is often hiding a deeper problem, such as data leakage, so it's a good idea to be skeptical of very high scores or to be skeptical of model results in general until the model has been thoroughly validated. This is a toy model. However, what this shows is that these embeddings can be created using a graph and that a supervised ML model can use these embeddings in prediction.

There's a nasty downside to using these embeddings with models, though. You lose all interpretability. With our hand-crafted training data, as shown earlier in this chapter, we could see which features the model found to be most useful in making predictions. Let's inspect the importances with these embeddings:

```
importances = pd.DataFrame(clf.feature_importances_, index=X_
train.columns)
importances.columns = ['importance']
importances.sort_values('importance', ascending=False,
inplace=True)
importances[0:10].plot.barh(figsize=(10,4)).invert_yaxis()
```

We get this output:

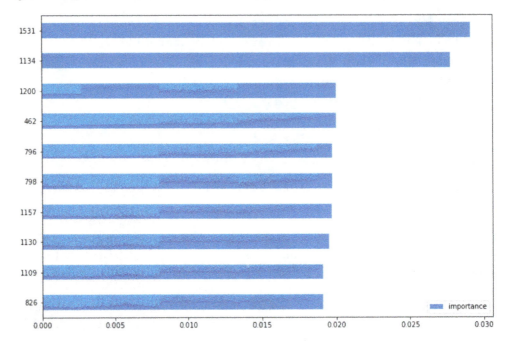

Figure 11.23 – FEATHER embedding feature importance

Wonderful! Embedding 1531 was found to be slightly more useful than 1134, but both of these were found to be quite a bit more useful than the other embeddings! Excellent! This is a total loss of interpretability, but the embeddings do work. If you just want to go from graph to ML, this approach will work, but you end up with a black-box model, no matter which model you use for prediction.

OK, for the rest of the models, I'm going to go a lot faster. We're going to reuse a lot of this code, I'm just going to do less visualization and give less code so that this chapter doesn't end up being 100 pages long.

NodeSketch

The next algorithm we will look at is **NodeSketch**, and you can learn about it at `https://karateclub.readthedocs.io/en/latest/modules/root.html#karateclub.node_embedding.neighbourhood.nodesketch.NodeSketch`.

In Jupyter, if you *Shift + Tab* into the model instantiation code, you can read about it:

An implementation of "NodeSketch" <https://exascale.info/assets/pdf/
yang2019nodesketch.pdf>

from the KDD '19 paper "NodeSketch: Highly-Efficient Graph Embeddings

via Recursive Sketching". The procedure starts by sketching the self-loop-augmented
adjacency matrix of the graph to output low-order node embeddings, and then
recursively generates k-order node embeddings based on the self-loop-augmented
adjacency matrix and (k-1)-order node embeddings.

Like FEATHER, NodeSketch also creates node embeddings:

1. Let's use the model and do the visualization in one shot:

    ```
    from karateclub.node_embedding.neighbourhood.nodesketch
    import NodeSketch
    model = NodeSketch()
    model.fit(G)
    embeddings = model.get_embedding()
    title = 'Les Miserables Character Similarity
    (NodeSketch)'
    draw_clustering(embeddings, nodes, title)
    ```

The following graph is the result:

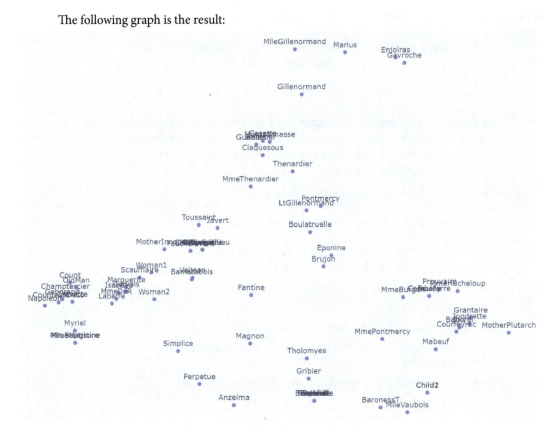

Figure 11.24 – NodeSketch embeddings

2. As before, this visualization is interactive and you can zoom in on clusters of nodes for closer inspection. Let's look at a few nodes that were found to be similar.

    First, **Eponine**:

```
node = 'Eponine'
G_ego = nx.ego_graph(G_named, node)
draw_graph(G_ego, show_names=True, node_size=3)
```

You can see the result in *Figure 11.25.*

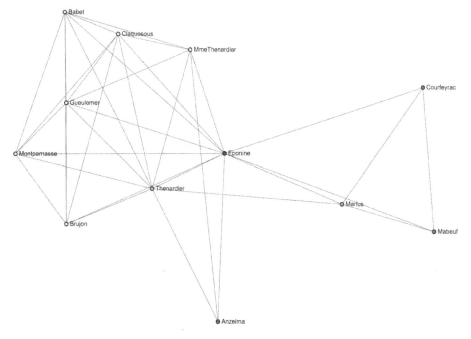

Figure 11.25 – Eponine ego network

3. Next, **Brujon**:

```
node = 'Brujon'
G_ego = nx.ego_graph(G_named, node)
draw_graph(G_ego, show_names=True, node_size=3)
```

Shown in *Figure 11.26*.

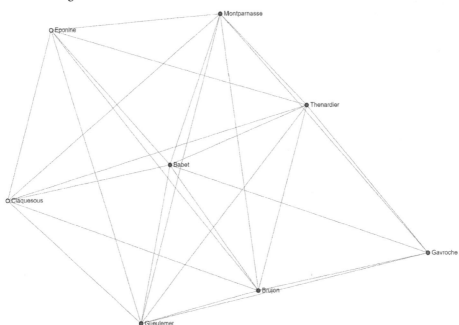

Figure 11.26 – Brujon ego network

Upon inspection, the ego networks look quite different, but the two nodes seem to have about the same number of connections, and they are part of a pretty well-connected community. I'm satisfied that these two nodes are pretty similar in structure and placement. Both nodes are also part of the same community.

4.   What do the embeddings look like?

```
eb_df = pd.DataFrame(embeddings, index=nodes)
eb_df['label'] = clf_df['label']
eb_df.head(10)
```

This will show our DataFrame.

	0	1	2	3	4	5	6	7	8	9	...	23	24	25	26	27	28	29	30	31	label
Napoleon	1	1	0	1	0	0	0	1	1	0	...	1	0	1	0	0	1	0	1	1	0
Myriel	1	3	0	10	5	5	9	1	2	2	...	2	6	10	10	4	9	10	4	7	0
MlleBaptistine	1	3	3	10	2	10	2	1	2	2	...	2	10	10	10	3	3	10	10	1	0
MmeMagloire	1	3	3	10	2	10	2	1	2	2	...	2	10	10	10	3	3	10	10	1	0
CountessDeLo	1	4	1	4	1	4	1	1	1	4	...	1	4	1	4	4	4	1	4	1	0
Geborand	1	1	5	1	5	5	5	1	1	5	...	1	5	1	5	5	5	1	1	1	0
Champtercier	1	1	6	1	6	6	1	1	6	6	...	1	6	1	1	6	6	1	1	1	0
Cravatte	1	1	7	7	7	7	7	1	7	7	...	1	1	1	7	7	7	7	1	7	0
Count	1	1	8	1	8	8	1	1	8	1	...	1	8	8	1	8	8	1	1	1	0
OldMan	1	9	9	9	1	9	9	1	9	9	...	1	1	9	1	9	9	1	1	1	0

10 rows × 33 columns

Figure 11.27 – NodeSketch embeddings DataFrame

Wow, this is a much simpler dataset than what FEATHER produced. **32** features rather than 1,750. Also, note that the values in the embeddings are integers rather than floats. How well is Random Forest able to make predictions with this training data?

```
X_cols = [c for c in eb_df.columns if c != 'label']
X = eb_df[X_cols]
y = eb_df['label']
X_train, X_test, y_train, y_test = train_test_split(X, y,
random_state=1337, test_size=0.3)
clf = RandomForestClassifier(random_state=1337)
clf.fit(X_train, y_train)
train_acc = clf.score(X_train, y_train)
test_acc = clf.score(X_test, y_test)
print('train accuracy: {}\ntest accuracy: {}'.
format(train_acc, test_acc))
```

If we run the code, we get these results:

```
train accuracy: 0.9811320754716981
test accuracy: 1.0
```

The model was able to predict with 98% accuracy on the training data and with 100% accuracy on the test data. Again, never ever test a model that gives 100% accuracy. But this still shows that the model is able to use the embeddings.

5. What features did it find important?

```
importances = pd.DataFrame(clf.feature_importances_,
index=X_train.columns)
importances.columns = ['importance']
importances.sort_values('importance', ascending=False,
inplace=True)
importances[0:10].plot.barh(figsize=(10,4)).invert_
yaxis()
```

This results in *Figure 11.28*.

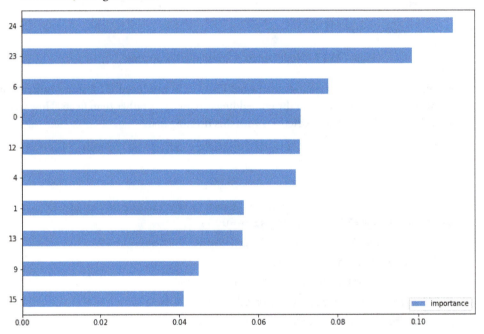

Figure 11.28 – NodeSketch embedding feature importance

Great. As I showed before, the use of these embeddings has turned Random Forest into a black-box model that we cannot get any model interpretability for. We know that the model found features **24** and **23** to be most useful, but we have no idea why. I won't be showing feature importances after this. You get the point.

This is a cool model, and it creates simpler embeddings by default than FEATHER. Random Forest did well with both models' embeddings, and we can't say which is better without a lot more experimentation, which is outside of the scope of this chapter. Have fun experimenting!

RandNE

Next up is **RandNE**, which claims to be useful for "billion-scale network embeddings." This means that this is useful for networks with either billions of nodes or billions of edges. It's a claim that would make this model useful for large real-world networks. You can read the documentation at `https://karateclub.readthedocs.io/en/latest/modules/root.html#karateclub.node_embedding.neighbourhood.randne.RandNE`.

In Jupyter, if you *Shift + Tab* into the model instantiation code, you can read about it:

> *An implementation of "RandNE" <https://zw-zhang.github.io/files/2018_ICDM_RandNE.pdf> from the ICDM '18 paper "Billion-scale Network Embedding with Iterative Random Projection". The procedure uses normalized adjacency matrix based smoothing on an orthogonalized random normally generate base node embedding matrix.*

1. Once again, let's generate the embeddings and do the visualization in one shot:

    ```
    from karateclub.node_embedding.neighbourhood.randne
    import RandNE
    model = RandNE()
    model.fit(G)
    embeddings = model.get_embedding()
    title = 'Les Miserables Character Similarity (RandNE)'
    draw_clustering(embeddings, nodes, title)
    ```

The output is the following graph:

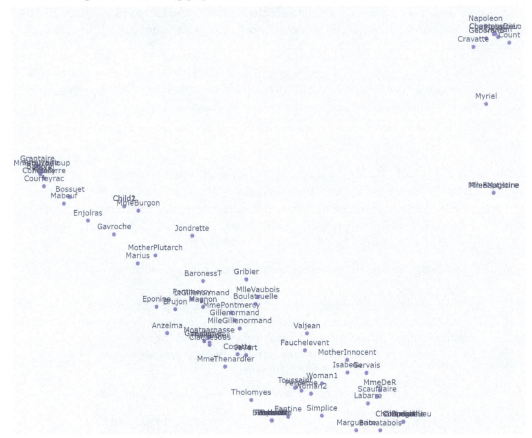

Figure 11.29 – RandNE embeddings

2. Right away, you can see that this scatterplot looks very different from both FEATHER and NodeSketch. Let's take a look at the ego networks for `Marius` and `MotherPlutarch`, two nodes that have been found to be similar:

```
node = 'Marius'
G_ego = nx.ego_graph(G_named, node)
draw_graph(G_ego, show_names=True, node_size=3)
```

We get a network output:

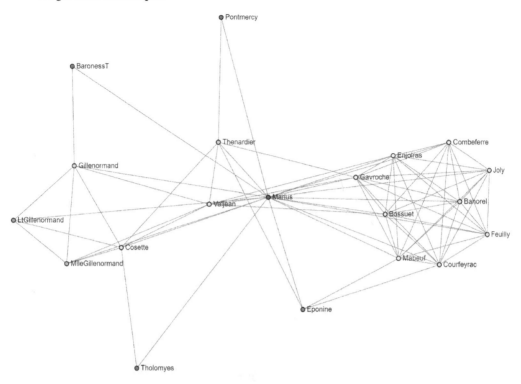

Figure 11.30 – Marius ego network

3. Next, **MotherPlutarch**:

```
node = 'MotherPlutarch'
G_ego = nx.ego_graph(G_named, node)
draw_graph(G_ego, show_names=True, node_size=3)
```

And the network is as follows:

Figure 11.31 – MotherPlutarch ego network

Wow, these ego networks are so different, and so are the nodes. **Marius** is a well-connected node, and **MotherPlutarch** has a single edge with another node. These are two very different nodes, and the embeddings found them to be similar. However, it could be due to the PCA step for the scatter plot visualization, so please don't be too quick to judge RandNE from this one example. Check out some of the other similar nodes. I will leave this up to you, for your own practice and learning.

What do the embeddings look like?

```
eb_df = pd.DataFrame(embeddings, index=nodes)
eb_df['label'] = clf_df['label']
eb_df.head(10)
```

This will show our embeddings.

	0	1	2	3	4	5	6	7	8	9	...	68	69	70
Napoleon	-0.948999	-0.370539	1.019771	-1.546844	1.402960	1.120856	-0.568228	-1.519486	0.948075	-1.082952	...	1.906159	2.644384	0.218053
Myriel	-0.610640	0.090700	0.049499	-1.010176	0.796567	0.448242	1.176782	0.142189	1.695997	-0.956446	...	0.238334	1.600401	-0.011550
MlleBaptistine	-0.817743	-0.554192	0.195265	-1.858893	1.151244	-0.231442	2.786735	0.482611	1.710762	0.601138	...	1.414692	0.623162	0.254414
MmeMagloire	-0.817743	-0.554192	0.195265	-1.858893	1.151244	-0.231442	2.786735	0.482611	1.710762	0.601138	...	1.414692	0.623162	0.254414
CountessDeLo	-0.216952	-1.242731	2.450365	-0.852710	0.944617	0.507591	0.425176	-1.436033	2.486548	-1.007876	...	0.114750	1.392726	0.630268
Geborand	-0.120592	1.374095	1.992590	0.611790	1.330882	1.169039	2.547460	-1.041690	1.818367	-1.134910	...	1.609993	1.391578	0.850674
Champtercier	0.169160	1.485793	0.963990	-0.051599	0.582123	2.055235	0.132082	0.604906	1.326603	-0.193353	...	-0.849449	0.915424	-0.090989
Cravatte	-0.288429	-0.158740	0.461061	0.226257	0.898541	0.576785	0.636480	-2.532547	0.761615	-0.566386	...	-0.349996	3.087163	-1.298550
Count	-2.173779	0.361887	1.540971	-0.698783	0.742119	0.261203	0.860602	1.204151	0.853675	-1.458796	...	0.860835	1.060284	-1.009853
OldMan	0.181095	-1.714148	0.102184	-0.943174	1.800889	3.403846	0.165649	0.168705	2.112665	-2.533567	...	0.400448	0.939491	-1.681064

10 rows × 78 columns

Figure 11.32 – RandNE embeddings DataFrame

The embeddings ended up being 77 features, so this creates simpler embeddings by default than FEATHER. NodeSketch created 32 features, in comparison.

4. How well is Random Forest able to use the embeddings?

```
X_cols = [c for c in eb_df.columns if c != 'label']
X = eb_df[X_cols]
y = eb_df['label']
X_train, X_test, y_train, y_test = train_test_split(X, y,
random_state=1337, test_size=0.3)
clf = RandomForestClassifier(random_state=1337)
clf.fit(X_train, y_train)
train_acc = clf.score(X_train, y_train)
test_acc = clf.score(X_test, y_test)
print('train accuracy: {}\ntest accuracy: {}'.
format(train_acc, test_acc))
```

If we run this code, we get these results:

```
train accuracy: 0.9811320754716981
test accuracy: 0.9166666666666666
```

The model was able to predict on the test set with 98.1% accuracy, and 91.7% accuracy on the test set. This is worse than with FEATHER and NodeSketch embeddings, but it could be a fluke. I wouldn't trust these results with so little training data. The model was able to successfully use the embeddings as training data. However, as before, if you inspect the feature importances of the embeddings, you will not be able to interpret the results.

Other models

These are not the only three models that Karate Club has available for creating node embeddings. Here are two more. You can experiment with them the same way that we did with FEATHER, NodeSketch, and RandNE. The results with Random Forest for all of the embedding models were about the same. They can all be useful. I recommend that you get curious about Karate Club and start investigating what it has available.

These models do the same thing, but the implementation is different. Their implementations are very interesting. I recommend that you read the papers that were written about the approaches. You can see these as an evolution for creating node embeddings.

GraRep

GraRep is another model we can use. You can find the documentation here: `https://karateclub.readthedocs.io/en/latest/modules/root.html#karateclub.node_embedding.neighbourhood.grarep.GraRep`:

```
from karateclub.node_embedding.neighbourhood.grarep import
GraRep
model = GraRep()
model.fit(G)
embeddings = model.get_embedding()
```

DeepWalk

DeepWalk is another possible model we can use: `https://karateclub.readthedocs.io/en/latest/modules/root.html#karateclub.node_embedding.neighbourhood.deepwalk.DeepWalk`:

```
from karateclub.node_embedding.neighbourhood.deepwalk import
DeepWalk
model = DeepWalk()
model.fit(G)
embeddings = model.get_embedding()
```

Now that we have several options for creating graph embeddings, let's use them in supervised ML for classification.

Using embeddings in supervised ML

Alright! We've made it through some really fun hands-on work involving network construction, community detection, and both unsupervised and supervised ML; done some egocentric network visualization; and inspected the results of the use of different embeddings. This chapter really brought everything together. I hope you enjoyed the hands-on work as much as I did, and I hope you found it useful and informative. Before concluding this chapter, I want to go over the pros and cons of using embeddings the way that we have.

Please also keep in mind that there are many other classification models we could have tested with, not just Random Forest. You can use these embeddings in a neural network if you want, or you could test them with logistic regression. Use what you learned here and go have as much fun as possible while learning.

Pros and cons

Let's discuss the pros and cons of using these embeddings. First, let's start with the cons. I've already mentioned this a few times in this chapter, so I'll just repeat it one last time: if you use these embeddings, no matter how interpretable a classification model is, you lose all interpretability. No matter what, you now have a black-box model, for better or for worse. If someone asks you why your model is predicting a certain way, you'll just have to shrug and say it's magic. You lost the ability to inspect importances when you went with embeddings. It's gone. The way back is to use hand-crafted network training data like we created at the beginning of this chapter and in the previous chapter, but that requires knowledge of network science, which is probably why some people are happy to just use these embeddings. This leads to the benefit of these embeddings, the pros.

The benefit is that creating and using these embeddings is much easier and much faster than creating your own training data. You have to know about network science to know what centralities, clustering coefficients, and connected components are. You don't have to know anything about network science to run this code:

```
from karateclub.node_embedding.neighbourhood.grarep import
GraRep
model = GraRep()
model.fit(G)
embeddings = model.get_embedding()
```

It's a problem when people blindly use stuff in data science, but it happens all the time. I am not excusing it. I am stating that this happens all over the place, and Karate Club's embeddings give you a shortcut to not really needing to know anything about networks to use graph data in classification. I think that's a problem, but it doesn't just happen with graphs. It happens in NLP and ML in general, all the time.

Loss of explainability and insights

The worst part about using these embeddings is you lose all model explainability and insights. Personally, building models doesn't excite me. I get excited about the insights I can pull from predictions and from learned importances. I get excited about understanding the behaviors that the models picked up on. With embeddings, that's gone. I've thrown interpretability in the trash in the hope of quickly creating an effective model.

This is the same problem I have with **PCA**. If you use it for dimension reduction, you lose all interpretability. I hope you have done the science first before deciding to use either PCA or graph embeddings. Otherwise, it's data alchemy, not data science.

An easier workflow for classification and clustering

It's not all bad, though. If you find that one of these types of embeddings is reliable and very high-quality, then you do have a shortcut to classification and clustering, so long as you don't need the interpretability. You can go from a graph to classification or clustering in minutes rather than hours. That's a huge speedup compared to hand-crafting training data from graph features. So, if you just want to see whether a graph can be useful for predicting something, this is one definite shortcut to building a prototype.

It's all pros and cons and use cases. If you need interpretability, you won't get it here. If you need to move fast, this can help. And it's very likely that there are insights that can be harvested from embeddings after learning more about them. I have seen that come true as well. Sometimes there are insights that you can find – it just takes an indirect approach to get to them, and hopefully, this book has given you some ideas.

Summary

I can't believe we've made it to this point. At the beginning of this book, this felt like an impossible task, and yet here we are. In order to do the hands-on exercises for this chapter, we've used what we learned in the previous chapters. I hope I have shown you how networks can be useful, and how to work with them.

At the beginning of this book, I set out to write a practical hands-on book that would be code-heavy, not math-heavy. There are tons of network analysis books out there that have an emphasis on math but do not show actual implementation very well, or at all. I hope this book has effectively bridged the gap, giving a new skill to coders, and showing social scientists programmatic ways to take their network analysis to new heights. Thank you so much for reading this book!

Index

Packt.com

Subscribe to our online digital library for full access to over 7,000 books and videos, as well as industry leading tools to help you plan your personal development and advance your career. For more information, please visit our website.

Why subscribe?

- Spend less time learning and more time coding with practical eBooks and Videos from over 4,000 industry professionals

- Improve your learning with Skill Plans built especially for you

- Get a free eBook or video every month

- Fully searchable for easy access to vital information

- Copy and paste, print, and bookmark content

Did you know that Packt offers eBook versions of every book published, with PDF and ePub files available? You can upgrade to the eBook version at packt.com and as a print book customer, you are entitled to a discount on the eBook copy. Get in touch with us at customercare@packtpub.com for more details.

At www.packt.com, you can also read a collection of free technical articles, sign up for a range of free newsletters, and receive exclusive discounts and offers on Packt books and eBooks.

Other Books You May Enjoy

If you enjoyed this book, you may be interested in these other books by Packt:

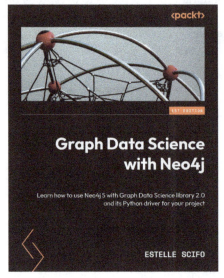

Graph Data Science with Neo4j

Estelle Scifo

ISBN: 9781804612743

- Use the Cypher query language to query graph databases such as Neo4j
- Build graph datasets from your own data and public knowledge graphs
- Make graph-specific predictions such as link prediction
- Explore the latest version of Neo4j to build a graph data science pipeline
- Run a scikit-learn prediction algorithm with graph data
- Train a predictive embedding algorithm in GDS and manage the model store

Modern Time Series Forecasting with Python

Manu Joseph

ISBN: 9781803246802

- Find out how to manipulate and visualize time series data like a pro
- Set strong baselines with popular models such as ARIMA
- Discover how time series forecasting can be cast as regression
- Engineer features for machine learning models for forecasting
- Explore the exciting world of ensembling and stacking models
- Get to grips with the global forecasting paradigm
- Understand and apply state-of-the-art DL models such as N-BEATS and Autoformer
- Explore multi-step forecasting and cross-validation strategies

Packt is searching for authors like you

If you're interested in becoming an author for Packt, please visit `authors.packtpub.com` and apply today. We have worked with thousands of developers and tech professionals, just like you, to help them share their insight with the global tech community. You can make a general application, apply for a specific hot topic that we are recruiting an author for, or submit your own idea.

Share Your Thoughts

Now you've finished *Network Science with Python*, we'd love to hear your thoughts! Scan the QR code below to go straight to the Amazon review page for this book and share your feedback or leave a review on the site that you purchased it from.

`https://packt.link/r/1-801-07369-4`

Your review is important to us and the tech community and will help us make sure we're delivering excellent quality content.

Download a free PDF copy of this book

Thanks for purchasing this book!

Do you like to read on the go but are unable to carry your print books everywhere?

Is your eBook purchase not compatible with the device of your choice?

Don't worry, now with every Packt book you get a DRM-free PDF version of that book at no cost.

Read anywhere, any place, on any device. Search, copy, and paste code from your favorite technical books directly into your application.

The perks don't stop there, you can get exclusive access to discounts, newsletters, and great free content in your inbox daily

Follow these simple steps to get the benefits:

1. Scan the QR code or visit the link below

https://packt.link/free-ebook/9781801073691

2. Submit your proof of purchase
3. That's it! We'll send your free PDF and other benefits to your email directly